Contents

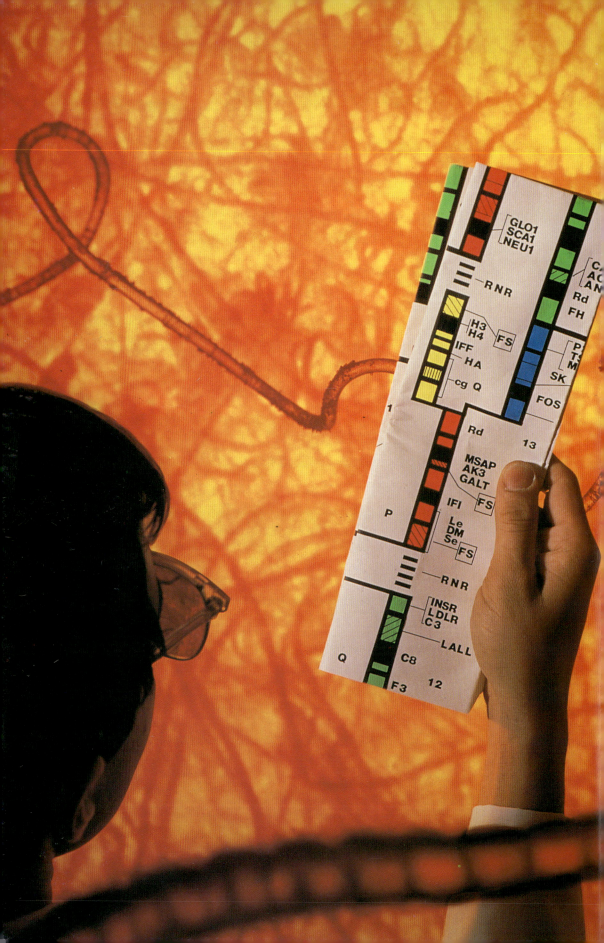

Library   Still in print 5/07

S C I

BIO

00062368

00

Published by the Press Syndicate of the University of Cambridge
The Pitt Building, Trumpington Street, Cambridge CB2 1RP, UK
40 West 20th Street, New York, NY 10011–4211, USA
10 Stamford Road, Oakleigh, Melbourne 3166, Australia

Printed in Hong Kong by Colorcraft

National Library of Australia cataloguing in publication data
Morris, Beryl.
Biotechnology.
Includes index.
1. Biotechnology – Juvenile literature. I. CSIRO. II. Title.
(Series: Science and our future).
660.6

Library of Congress cataloguing in publication data
Morris, Beryl.
       p.   cm.– (Science and our future)
Includes index.
1. Biotechnology – Juvenile literature. [1. Biotechnology.
2. Genetic engineering.]
I. Title. II. Series.
TP248.218.M67 1994
660'.6–dc20   93– 41597
                                                        CIP
                                                        AC

A catalogue record for this book is available from the British Library.

ISBN 0 521 43785 7 Paperback

Notice to Teachers

Acknowledgements

I am indebted to Dr John Watson of the CSIRO Division of Plant Industry, who read the
manuscript. Dr Nigel Scott of the Division of Horticulture provided the latest information from
his grapevine DNA fingerprinting work. There are many photographers who have generously
allowed their material to be reproduced: Dr M. Ryder (CSIRO Division of Soils), G. Brown
(CSIRO Division of Plant Industry) and E. Lawton, Dr N. Scott and E. Dare (CSIRO Division
of Horticulture). Gary Underwood of Cambridge University Press has been consistently
supportive and Mitch O'Toole contributed generously to the student questions. There are many
Australian research teams and individuals working on biotechnology. While only some are
mentioned specifically in this book, without their collective work, this book would not have
been possible.

Grateful acknowledgement is made to the following for their assistance in both providing
suitable material and granting Cambridge University Press permission to reproduce it in this
text. Photographs (pp. iv, 10, 66, 78), Stock Photos; photographs (pp. 20, 38, 50), Horizon Photo
Library; photograph (Fig. 4.2), Kraft Foods Limited; photograph (Fig. 3.10), Commonwealth
Serum Laboratory; photograph (Fig. 5.12), Fredy Mercay/A.N.T. Photo Library; photograph
(Fig. 1.3), Australian Government Information Service. Every effort has been made to trace and
acknowledge copyright but in some cases this has not been possible. Cambridge University
Press would welcome any information that would redress this situation.

Chapter 1
Biotechnology

People have been brewing beer with yeast, making cheese with bacteria and cross-breeding plants and animals for centuries. Although they did not understand how the processes worked, through trial and error they learned to control them.

We use animals, plants, yeasts, viruses, fungi, moulds and bacteria to produce useful materials like food, medicine and chemicals. When we do this, it is called biotechnology. Biotechnology is all about industrial processes based on biological systems. The biological systems can involve naturally occurring micro-organisms, genetically engineered micro-organisms, or isolated cells of plants and animals. Biotechnology also includes the genetic manipulation of cells to produce new strains.

In modern biotechnology, we aim to make a living cell perform a task in a predictable and controllable way. Whether a living cell will perform that task depends on how just four types of chemicals are arranged along the large ladder-like molecule known as DNA. Particular groups of these four chemicals are called genes. We have been choosing

Figure 1.1 Hand pollinating a citrus flower

1

desirable genes for centuries — by selective breeding. Results include the merino sheep breed, rust-resistant wheat, and bacteria and fungi that make high levels of antibiotics.

Discoveries in the early 1970s made it possible to transfer genetic material between unrelated organisms. An example is the human gene for the hormone insulin. This gene has been introduced into bacteria. Large amounts of pure insulin are now available for diabetes sufferers.

Genetic engineering is adding or subtracting genes in an organism, or changing the way genes work, by non-sexual processes. Altering the genes of animals, plants and micro-organisms could help to fight infectious diseases, improve agricultural production and help reduce pollution.

Is there a risk with genetic engineering?

Genetic engineering has the potential to benefit humanity. It could increase production from crops and livestock or help prevent or fight diseases. However, many people are concerned that genetic engineering will change familiar species, making them dangerous pests or a threat to life, that genetic engineering will create a 'Frankenstein's monster'. Most scientists agree that there really is little chance of creating a Frankenstein's monster. For each crossing in selective breeding, hundreds or thousands of unknown genes are combined. Genetic engineering introduces only one or two well-studied genes into an organism causing predictable

changes. For this reason, genetic engineering is seen as a great improvement on the more random process of selective breeding.

Others worry more about the issues of commercial control or subtle manipulation than about the creation of a monster.

People, of course, differ in their view of the relationship between humanity and nature. This is one of the reasons why people have differing views of genetic engineering. Each of us has to decide for ourselves. Our opinions will probably take into account the question of whether we think the technology interferes with nature. Possible effects on the environment are important too. Most of us are health-conscious and would also be concerned about whether we should use the technology on humans. These concerns lead us to ask whether anyone, companies for instance, have the right to profit from genetically engineered organisms and who should have control over genetic engineering. We also have to consider the present effect of humanity and its technologies on the world's resources.

Ideally, two ideas should be considered before answering these questions.
- Basic philosophical concerns underlie many of the objections to genetic engineering.
- It is not 'genetic engineering' that is right or wrong, it is the way in which we use the technology.

Experience shows that once scientific knowledge is revealed it cannot be erased. Technical developments can help people's lives, or they may be harmful.

Should scientists be stopped from studying certain parts of the natural world simply because the results might be misused in the future? The results might also be put to a worthwhile use. It is better that the community becomes aware of the science and its various potential uses. The community will then decide what uses to allow. If the community is against genetic engineering, then it is likely that funds for further research would be difficult to obtain. The research would then stop.

Who is in charge?

North America and Europe have taken different approaches to the regulation of genetic engineering. In North America, the approach is called product-based regulation. In Europe it is called process-based.

In Canada and the USA, those assessing the risks of the technology have argued that regulations should look at the final product and its intended use. With this view, current systems of regulation can be used. Responsibility for regulating the product falls to the agency that also has responsibility for the equivalent naturally occurring product. In North America, industries have been very active in developing new products and carrying out field trials on transgenic plants.

In Europe, the regulators feel that the process of modifying an organism using genetic engineering introduces new types of risks. They therefore have new forms of regulation. The European Economic Community requires all member countries to have special laws to regulate both contained research and the release of genetically modified organisms.

Some European companies have moved their biotechnology activities out of Europe to avoid these regulations.

The first government approval for the release for field trial of a transgenic plant was made in the USA in 1986. Since then, the number of approvals has nearly doubled each year, and by late 1992 there had been 864 approvals granted at 1185 sites across 15 OECD countries. The USA and Canada accounted for 70 per cent of these approvals.

Release approvals have been granted for over 30 different crops; the five major crops are canola, potato, tomato, tobacco and corn. The main traits introduced into the approved plants were herbicide tolerance and virus, insect and disease resistance. Attention is now being given to quality traits of plants.

Biopharmaceuticals, mostly owned and produced by USA companies, are generating huge sales. In 1992, erythropoietin became the first product derived from genetic engineering to generate sales in excess of US$1 billion worldwide. Hepatitis B vaccine, human insulin, human growth hormone, a-interferon, and the white cell stimulator, G-CSF, all had sales in excess of US$500 million in 1992. Two-thirds of the Investigational New Drugs applications to the USA Food and Drug Authority are products derived from biotechnology, mostly genetic engineering.

Right from the beginning of genetic engineering research in Australia, in the early 1970s, it was recognised as a very powerful technology. Some of the scientists working with the technique became so concerned that it might be possible to make hazardous micro-organisms that they called for an investigation into the safety of the technique. Australians were not alone in this concern.

In 1975, molecular biologists from around the world, including Australia, met in the USA. Delegates asked the United States National Institutes of Health to develop guidelines for the safe conduct of genetic engineering experiments. These guidelines were distributed in 1976. The participants also asked for a halt to some types of research until the risks had been more fully identified.

Australia did not just rely on overseas countries to decide what was acceptable. The Australian Academy of Science, a group of our most respected scientists, set up a Committee on Recombinant DNA that drew up the first Australian guidelines for the technique in 1975. In 1981, the Australian Government's Department of Science established the Recombinant DNA Monitoring Committee. This group reported to its Minister in 1986 that monitoring of the technology up to date had been effective and would be likely to remain so for at least the next five years and so should continue. In 1987, the Australian Government's Minister for Industry, Technology and Commerce established the Genetic Manipulation Advisory Committee (GMAC) to replace the Recombinant DNA Monitoring Committee. The Minister for Administrative Services took responsibility for GMAC in 1988. GMAC had its first meeting later that year.

The function of GMAC was to review proposals for genetic manipulation work in Australia to identify and manage any risks associated with the organisms that resulted from the technique. GMAC also advised the Minister on matters about regulating the technology.

Any institution or organisation conducting genetic engineering, or involved with genetically modified organisms, was expected to follow the GMAC guidelines. This meant the institution had to set up an Institutional Biosafety Committee to assess and review proposals for genetic engineering and to submit them to GMAC for assessment to ensure that they met the guidelines. Research institutions are also required to provide the resources and facilities for safe work and to ensure their staff were adequately trained and supervised. For planned release proposals, GMAC submitted its advice to the state and federal agencies that may have a legal jurisdiction over the proposed activity.

An example of this process is as follows. CSIRO's Division of Plant Industry submitted a proposal for the release of potato plants containing genetic material from potato leafroll virus which makes the plants resistant to the virus. The project was being supervised by the Institutional Biosafety Committee of the Queensland Department of Primary Industries. GMAC considered that there were no significant human health or environmental hazards from the trial. Among the recommendations, GMAC asked that the field trial be extended one year and that all plants in the trial be removed by hand at the end of the trial.

Institutions involved in the technology are legally liable for the work they do. It is also likely that in future, researchers will be liable if they release genetically modified organisms without proper approval. Approval will reduce their liability. Those who give approval will not be liable unless they have failed to do their jobs. New laws on product liability are being introduced that will also cover genetically modified products.

The House of Representatives Standing Committee on Industry, Science and Technology started a public inquiry into regulation of genetically modified organisms in June 1990. The Standing Committee had to identify issues associated with the creation and release of genetically modified organisms and advise on any new legislation needed. The inquiry was an attempt to reconcile the concerns of the public with the desire of the biotechnology industry to 'fast track' their new products. In February 1992, the Standing Committee released a report of its findings called 'Genetic manipulation: the threat or the glory?'

Figure 1.2 Researchers working on genetic engineering experiments follow strict safety procedures

Figure 1.3 The House of Representatives, Canberra

The Standing Committee did not interpret community opinion as demanding the outlawing of genetic manipulation and argued that genetic manipulation techniques are worth pursuing, even if not all the claimed benefits come about. It felt that the main concern was that the guidelines for research have no legal force.

Some lobby groups criticised the report when it was released, claiming that it did not recommend a way for the public to be informed of proposed releases of genetically modified organisms. The critics also pointed out that the same government department that approves research and releases, the Department of Industry, Technology and Regional Development, also provides money for genetic engineering research and development programs and produces publications promoting gene technology. They claimed that this may involve a conflict of interest. In 1988, responsibility for regulating genetic engineering transferred from this department to a more neutral agency — the Department of Administrative Services — because of this intense debate.

A total of $5.9 million over four years was approved for a new Genetic Manipulation Authority (GMA) in 1992. This body is to assess and approve proposals for the release of genetically modified organisms following the House of Representatives Standing Committee's recommendations.

GMA is to be a statutory authority responsible to the Minister for Industry, Technology and Regional Development and operated with GMAC. GMAC is to be renamed the Genetic Manipulation Research Committee (GMRC) and will once again be in the Department of Industry, Technology and Regional Development.

One of the biggest issues with genetically manipulated organisms is the question of the right to own an organism. Patents are a way of giving an incentive for development of new products. For publicly funded research, patents are a way of getting money back. Patenting living organisms is allowed under Australia's patent laws and the House of Representatives Standing Committee has recommended that the patent periods for products of genetically manipulated organisms be extended beyond the normal period of 16 years if they have had to undergo extensive testing. Those objecting to patents for living organisms claim it downgrades living things. The Standing Committee, however, felt that the patent system is not the best way to tackle objections based on aesthetics, biodiversity and so on.

For instance, the patent system does not reduce the need to consider animal welfare. Animal ethics groups in research institutions and animal welfare groups in state and local governments are in the best position to discuss most of the arguments about animal rights.

The House of Representatives Standing Committee's report does say that the public has a right to be informed if products contain or are produced by genetically modified organisms. Labels on products could give this information but some substances are not (at present) under strict product regulations. Nutritional additives are an example. The Standing Committee did make the point that genetic engineering does not in itself make a product hazardous. However, the method of manufacture could be relevant and so labelling should be looked at product-by-product. The report recommended that labels state the method of manufacture for some classes of foods. The report also suggested broadening product liability laws.

Genetic engineering: its effect on society

The 1993 Nobel Prizes in both medicine and chemistry were for discoveries in the field of gene technology. Forty scientists have now received Nobel Prizes for work that has contributed to the development of gene technology.

Gene technology has social and economic significance for Australia. Genetic engineering will affect our everyday lives, especially in health care, food production and the environment. The interest and rapid developments that are occurring around the world show that genetic engineering will be a major factor in determining the competitiveness of our agriculture and value-adding food-processing industries. Its applications in medicine will have benefits for our future health care.

If Australia does not participate in the technology, we will be left behind. Some say that we can get an idea of what the effect will be if we imagine that Australia chose not to be involved in computer technology in the 1960s. In the case of genetic engineering, we will miss the opportunity to provide better health care and a sustainable environment for our country. By participating now in the latest techniques, we will capture the benefits for Australia rather than having to buy them from overseas.

Genetic engineering is considered to be an environmentally friendly technique. The United Nations Conference on Environment and Development saw gene technology as an important tool for enhancing the protection of the environment and sustainable agriculture.

Australia's industry is well placed to take advantage of genetic engineering. We are very good at research and in industries such as agriculture and pharmaceuticals, and we are willing to take advantage of international markets.

The potential barriers to Australia's use of the technology include public perception and regulation. At present there is some concern about the possibility of misuse or adverse effects from genetic engineering. The technology is difficult to understand. Researchers are concerned that if genetic engineering is misunderstood, governments will impose regulations so strict that further work may be impossible. Scientists see that they need to explain the technology and take part in the debate about the technology so that companies, scientists and the public can understand each other.

Learning about biotechnology is important. The incorporation of biotechnology into the school curriculum is one way in which governments are ensuring that the information and tools needed to evaluate gene technologies are being given to the people who will be making many of the decisions about the technology's future.

1. Find the following words in this chapter. What does each word mean?
   - biotechnology ● patent
   - predictable ● biodiversity
   - liability ● regulated
   - genetic manipulation
2. Each paragraph in a well written passage will contain a single sentence which gives its main idea. The rest of the paragraph develops or supports the idea expressed in this key sentence. Write down the key sentence from each paragraph in this chapter. The list of key sentences is a crude summary of the chapter.
3. 'Biotechnology' is a new word. Discuss whether the thing it names is also new.
4. What are some of the risks of biotechnology?
5. How are the risks involved in biotechnology reduced?
6. How is biotechnology regulated in Australia?
7. Use a time line to show the names and locations of the changing committees that have overseen genetic engineering research in Australia.
8. This chapter has taken the position that genetic engineering is neither right nor wrong. Discuss this position and explain why you accept or reject it.

9. Summarise the arguments for and against genetic engineering.

10. Discuss the different interests and roles of science, industry, government and the wider community in the continuing debate about genetic engineering.

11. What is a 'lobby group'? Why are they called this? Do you think that lobby groups should have any role in decisions about things like genetic engineering? Give reasons for your answer.

12. Paragraphs are often tied together by 'pointer words'. Pointer words are words which refer to other parts of the paragraph. They are small words such as 'he', 'she', 'it', 'they', 'this', 'each', 'who' or 'that'. Turn back to the first page of this chapter. Take a pencil and put a circle around each pointer word. Draw an arrow from the circle to the words to which the pointer refers. The first pointer is done for you below.

→ People have been brewing beer with yeast, making cheese with bacteria and cross-breeding plants and animals for centuries. Although they did not understand how the processes worked, through trial and error (they) learned to control them.

13. 'Connectives' are words that link ideas. Connectives are joining words which show the relationship between the ideas they connect. Choose words from the following list to join the pairs of sentences below.

- after
- although
- and
- as
- because
- but
- eventually
- furthermore
- meanwhile
- often
- or
- similary
- where
- which

Write the joined sentences. You may need to change some other words to combine the sentences. You may use the same word more than once and you can use more than one word from the list in each pair of sentences. The first pair is done for you.

Genetic engineering may be able to help people.
It could help to produce more food.

Genetic engineering may be able to help people as it could help to produce more food.

It could help stop disease.
Many people are worried that genetic engineering will change common animals and plants.

The changed species might be dangerous pests.
The changed species might be a danger to life.

The changed species might be like the monster in 'Frankenstein'.
Most scientists agree that there is little chance of making such monsters.

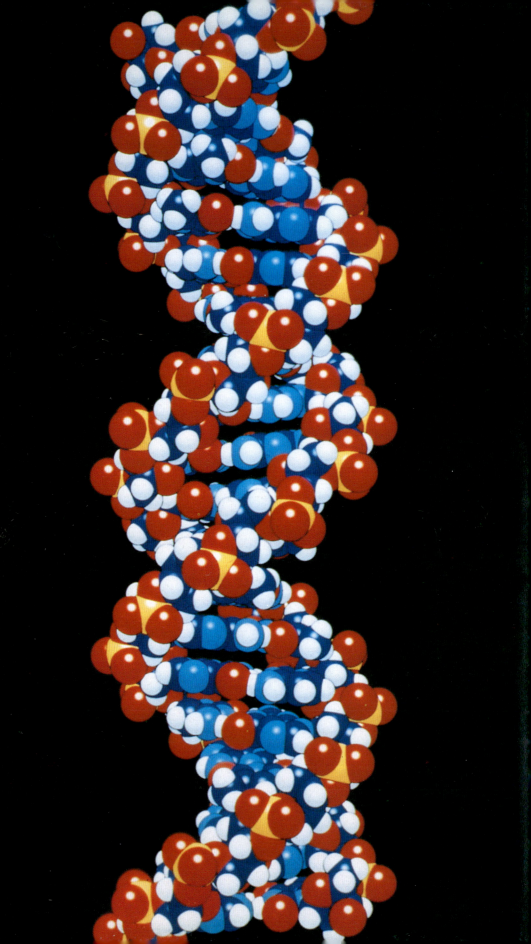

Chapter 2
Genetic engineering

The transfer of genes from one organism to another is a natural process, but in nature is restricted to members of the same species. Genetic engineering is the laboratory technique for changing the hereditary material of a living cell. The changed cell can then produce more or different chemicals, or perform completely new functions. With genetic engineering it is possible to introduce desirable new genes into plants, animals, bacteria and fungi in a precise, controlled and predictable way — a process that might otherwise take decades of systematic breeding and selection. It also means we have the potential to detect and control genetic diseases.

Genes

To understand genetic engineering, we need to know about genes. All cells in our bodies (except sperm and egg cells), contain 46 chromosomes in their nuclei. (Egg and sperm cells contain 23 chromosomes.) Each of us has inherited 23 of the chromosomes

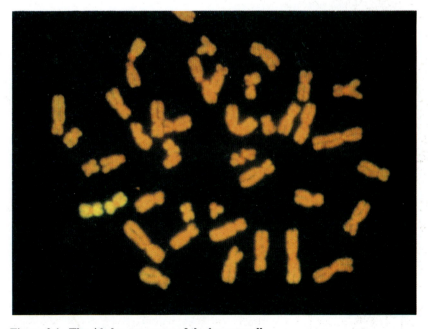

Figure 2.1 The 46 chromosomes of the human cell

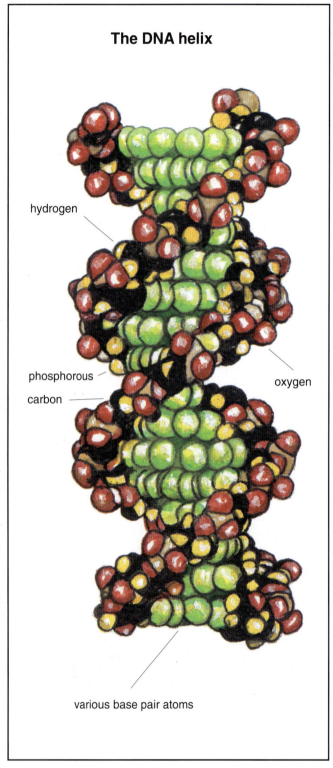

The DNA helix

hydrogen

phosphorous

carbon

oxygen

various base pair atoms

Figure 2.2 The DNA molecule is shaped like a twisted ladder

from our mother (through an egg cell) and 23 from our father (through a sperm cell).

Chromosomes are made of a molecule called DNA, which stands for deoxyribonucleic acid. DNA is tightly coiled and folded in chromosomes. It is a spiral molecule, called a helix, and looks like a twisted ladder. Pairs of nucleic acids, or 'bases', form the rungs. The bases are Guanine (G), Cytosine (C), Adenine (A) and Thymine (T). Because of the bases' chemical properties, A pairs only with T and C only with G. Sugar-phosphate units make the sides of the ladder. The sugar is deoxyribose.

The 46 chromosomes in each cell in our bodies contain all the information to make an entire human. The DNA molecule carries the information in code. The sequence of the four bases make the code. Each section of the code, or group of bases, carries the information for a particular protein. Such a group of bases is called a gene. The stretch of bases that form a gene may have from 1000 up to 20,000 bases. DNA carries all the genes to make all the proteins for an organism. You cannot see genes.

The genetic code describes the arrangement of the four bases, A, G, C and T. The linear sequence of the four bases in the DNA of every species contains the information for that particular organism.

Making proteins

The coded message of the gene has meaning only when translated in the form of a protein. Essentially, one

1. DNA or deoxyribose nucleic acid consists of a chain of nucleotides. There are four different nucleotides in DNA, which can be represented like this:

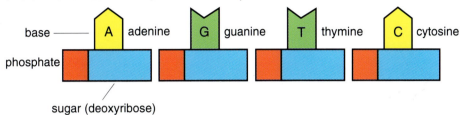

2. The nucleotides join together forming long chains:

3. The nucleotides on this chain form a code that can be written as:

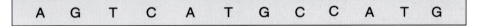

4. The bases on one chain can bind with the bases on another to form double-stranded DNA. Note that A always pairs with T and C with G.

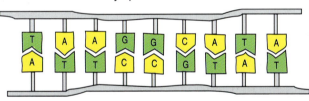

5. Two strands capable of pairing up are called complementary strands and form a long spiral molecule known as the double helix. It is made up of sugar-phosphate pairs.

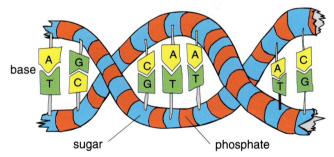

Figure 2.3 The 'rungs' of the DNA ladder are built of pairs of bases: Thymine (T), Adenine (A), Guanine (G), and Cytosine (C). Bases are also called nucleotides.

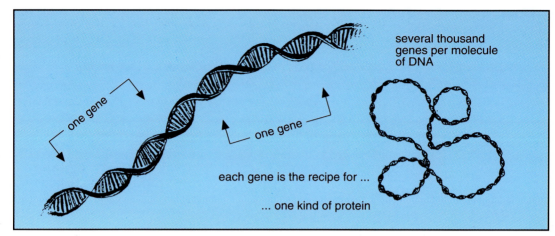

one gene

one gene

several thousand genes per molecule of DNA

each gene is the recipe for ...

... one kind of protein

Figure 2.4 Each section of the DNA code ia a group of bases called a gene. The coded message of the gene has meaning only when translated in the form of a pattern.

gene codes for one protein. Chains of amino acids make proteins and the sequence of the bases forming the code on the DNA molecule directs the order of the amino acids in a protein chain.

Codons, groups of three bases in the DNA molecule, make the genetic code. Each codon represents each amino acid. For example, the amino acid valine has the code GTG, AGC codes for serine while ACG signifies threonine. Since there are only four bases, there are 64 codons in all ($4 \times 4 \times 4$). There are only 20 amino acids to code for so there is coding information to spare. Three of the codons are 'stop' signals. Most amino acids can be coded for by more than one codon.

The code is linear and non-overlapping. This is why the gene sequence 'AGCTGTA' can be read in several ways — AGC TGT A, or

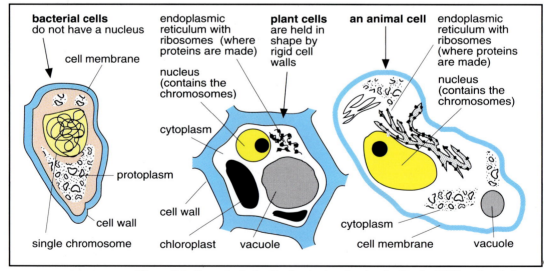

bacterial cells do not have a nucleus

cell membrane

protoplasm

cell wall

single chromosome

endoplasmic reticulum with ribosomes (where proteins are made)

nucleus (contains the chromosomes)

cytoplasm

cell wall

chloroplast   vacuole

plant cells are held in shape by rigid cell walls

an animal cell

endoplasmic reticulum with ribosomes (where proteins are made)

nucleus (contains the chromosomes)

cytoplasm

cell membrane   vacuole

Figure 2.5   Some typical cells. In plant and animal cells, the DNA molecule is found in the form of chromosomes in the cell nucleus. In bacterial cells, which do not have a nucleus, the DNA is freely coiled.

AG CTG TA or A GCT GTA, with each base being read only once. There are various 'start' signals coded in the DNA before each gene. These signals decide the code's translation.

The nucleus of animal and plant cells contain the genes. A thin membrane separates them from the cell's protein factories, the ribosomes, which are outside the nucleus in the cell's cytoplasm. A go-between molecule that can move through the membrane between the nucleus and the cytoplasm translates the genetic instructions into proteins. The go-between molecule is messenger ribonucleic acid, or mRNA. It is similar in structure to DNA but single-stranded and with the base Uracil (U) instead of Thymine (T).

An enzyme splits the DNA helix next to the gene coding for the required protein. One half of the DNA strand is then copied into an mRNA molecule. The DNA acts as a template and the mRNA copy of the gene, still in the nucleus of the cell, is called a transcript. After copying, the DNA reforms as before. The mRNA moves through the nuclear membrane to the cytoplasm. Within the cytoplasm, ribosomes attach themselves to the mRNA strand and the process of translating the coded information into an amino acid sequence begins.

This process is called 'translation' because the codons of mRNA are translated into the amino acids they encode. During translation, molecules called transfer ribonucleic acids (tRNA) bring amino acids to the appropriate place on the mRNA strand. Each tRNA has a triplet of bases, called an anticodon, which finds and joins up with its complementary codon on the mRNA strand.

These tRNAs ferry the correct amino acid to the ribosome and appropriate enzymes then link each one to the growing protein chain. As the ribosome moves along to the next codon on the mRNA, a new tRNA brings up the new amino acid.

When the protein chain is completed, it falls off the ribosome and twists into a predetermined shape. It is then ready for its next stage of processing and transport in the body.

Genetic engineering

Genetic exchange works constantly in all cells to rearrange chromosomes — such recombinations are an essential, normal feature of evolution. The relatively simple DNA rearrangements (cutting and recombining) performed in the test tube by molecular biologists are imitations of the natural processes that occur in the cell. The difference is that molecular biologists can join fragments of DNA that would not normally come together.

There are probably 100,000 genes in a typical plant or animal cell and biologists can now identify perhaps 500 of these, covering a small number of species. It is now possible to cut and splice genes from a DNA molecule and, in some cases, transplant them into the DNA of a new host. An animal, plant, bacterium or fungus that has had one of its genes altered by genetic engineering is called 'transgenic'.

Regardless of whether the target organism is a plant, animal, bacterium or fungus, this type of gene surgery can only occur in single cells. However, as the transformed cell divides and is organised into a new plant or animal, it passes on the altered genetic blueprint.

In genetic engineering, special proteins known as restriction enzymes cut the DNA molecule into short pieces. The required gene segment is removed and, using different enzymes, joined to the DNA strand of a so-called 'cloning vector'. The new DNA is then placed into organisms that can reproduce the required gene. Those organisms that have successfully taken up the new DNA are identified, isolated and grown to obtain useful quantities of the required gene. Finally, the new DNA is transferred into the target organism.

More than any other thing, the discovery of a tool for cutting DNA into fragments made genetic engineering possible. In 1969, the first enzyme to cut DNA at defined sites was isolated from bacteria. Many hundreds of such enzymes, called restriction endonucleases, that cut the DNA at different but known sites have now been recognised. The enzyme *Eco*RI, for example, cuts DNA at any site that has the sequence GAATTC and the enzyme *Hin*dIII cuts DNA at any site that reads AAGCTT. These enzymes normally protect the cell from any invading foreign DNA (from a virus, for example) by cutting it into small pieces, usually before it can replicate or recombine with the cell's own DNA.

Scientists use these enzymes to cut DNA into small fragments so they can:
- identify protein coding sequences;
- identify the function of encoded proteins; and
- insert a DNA fragment into a chosen organism.

If the DNA is from a simple organism, such as a virus or bacterium, an electrophoresis gel is used to separate the fragments according to length and isolate the required piece of DNA

In more complex organisms, such as humans, it is necessary to work with mRNA rather than DNA. An enzyme called reverse transcriptase reads the mRNA backwards to make a single-stranded complementary copy of the mRNA. Another DNA strand then

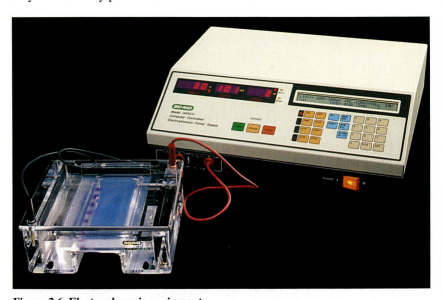

Figure 2.6 Electrophoresis equipment

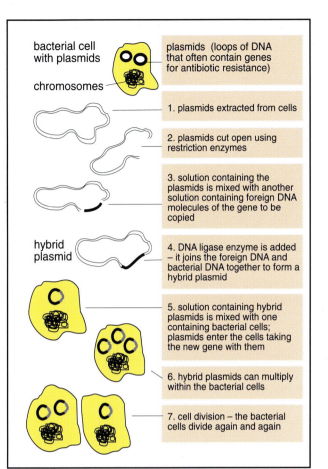

Figure 2.7   Using bacteria to make multiple copies of a gene from another organism

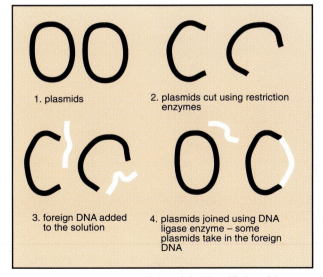

Figure 2.8   Not all plasmids contain the new gene

replaces the mRNA to make a double-stranded DNA molecule that is a copy of the DNA from which the original mRNA was made.

It is also possible to make a required gene in the laboratory simply by adding the bases together in the required sequence using an instrument known as an automated DNA synthesiser.

To make large amounts of the chosen gene, it is reproduced in a living cell. Circular pieces of DNA from bacteria, called plasmids, are generally used as the cloning vectors. Plasmids are easy to obtain because they are separate from the bacteria's normal chromosomal DNA. A restriction enzyme cuts the plasmid and another enzyme, DNA ligase, joins the chosen gene into the plasmid at the cut point. Circular recombinant plasmids result.

The recombinant plasmids are then introduced into bacteria. This is a hit and miss affair. Not all of the bacteria will take up a plasmid and not all plasmids will contain the new gene. Those that have taken up the new gene must be separated from those that have not. The commonly used plasmid cloning vectors contain a gene that makes the bacteria resistant to a particular antibiotic. Only those bacteria that have taken up a plasmid will be able to grow on a medium containing that particular antibiotic. Scientists identify which of the antibiotic resistant bacteria contain the required gene by marking it with a radioactive tag. A radioactive tag is also called a 'probe'.

Once there is a 'pure' culture of bacteria containing the required gene, they can be quickly reproduced in very large numbers. Where genetic engineering is being used to make large quantities of a desirable protein, such as insulin, the bacterial culture can now produce that protein.

Figure 2.9 A knowledge of the fine structure of genes and proteins is essential in genetic engineering. Computer graphics help us to understand the three dimensional structure of Neuraminidase, a protein molecule isolated from the outer surface of influenza virus.

If the desirable gene is to be introduced into another organism, say a plant, it has to be removed from the bacteria and introduced into the cells of the new host. Again, restriction enzymes separate the 'cloned' gene from the recombinant plasmids.

The cloned gene has to be inserted into the target species. For animals, the preferred technique is to micro-inject the gene into single-celled embryos. For plants, the most successful technique has been to use *Agrobacterium*. This bacterium naturally infects plant cells, inserting a small segment of its own DNA into the infected plant's DNA in the process . In the final step, tissue culture or embryo transfer techniques are used to grow the transformed cell into a new complete organism.

Gene Shears

As well as adding a gene to a cell, scientists can also stop an existing gene's instructions from being read. Gene Shears is the name CSIRO scientists have given to the mechanism of selectively switching off genes. Gene Shears are special RNA molecules called ribozymes. The technique works by the ribozymes cutting and destroying mRNA. This stops the transfer of genetic instructions from the gene to the ribosome protein factory and prevents the gene from expressing itself as a characteristic in an organism.

Initially the technology is likely to be used to introduce virus resistance into crop, horticultural and pasture plants. In the same way as Gene Shears can target mRNA, Gene Shears can also target viral RNA, destroying it as it enters a cell. Viruses are little more than shells of protein containing RNA molecules (although some also have DNA as their genetic material). They cannot reproduce themselves and need to parasitise the cells of living organisms to multiply, or replicate. A virus 'injects' its RNA into a cell, deceiving the cell into accepting the

Chapter 3
Health

One of the ways in which we find out how the world works is through science. Through science we are discovering more about the make up of human DNA and what each gene does. Moral issues of how to use the knowledge emerge as we learn more. We are moving into gene therapy, gene mapping, *in vitro* fertilisation, monoclonal antibodies and DNA fingerprinting. These are just some of the fields in which biotechnology is challenging sensitive moral and ethical boundaries in health and civil liberties.

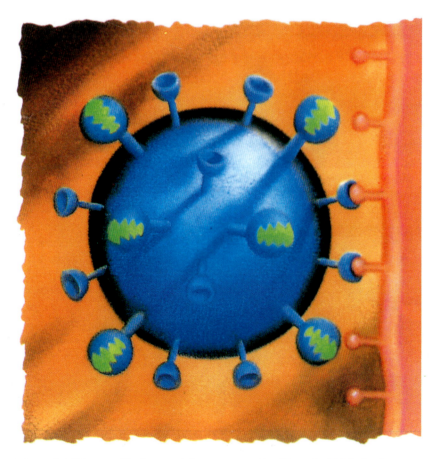

Figure 3.1 We are still a long way from conquering disease but biotechnology promises many breakthroughs

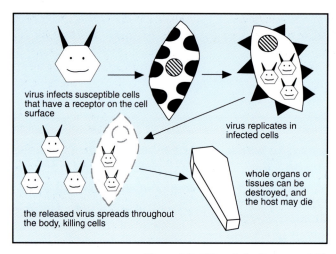

Figure 3.2   Virus infections may kill whole organs or tissues, killing the host.

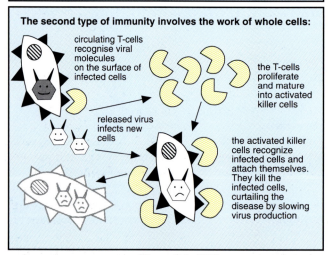

Figure 3.3   Different types of immunity

The immune defence system

Animals have an immune system. This system recognises and attacks organisms like bacteria, parasites and viruses when they invade our bodies. A healthy immune system limits the spread of infectious organisms through the body. It also helps the body to recover from any infectious diseases.

The immune system 'remembers' which micro-organism caused a disease. It deals with invasion much faster if the same micro-organism invades again. This provides us with immunity to diseases such as measles. When the measles virus invades the first time, our immune system takes a while to produce enough antibodies to fight off the virus. This is why we show signs of the infection, like fever and spots. The next time the measles virus invades, our immune system quickly fights it off before we develop any symptoms.

The immune system works because cells in our bodies, called B-cells, respond to antigens. Antigens are proteins found in or on the surface of the invading micro-organism. The B-cells produce other proteins, called antibodies. The antibodies bind to the antigen and so stop the invader. Each antibody binds to only one type of antigen. This means that the antibodies for the measles virus are useless in fighting the mumps virus. The antibody-coated micro-organism is finally eaten by macrophages, which are another special cell type in the immune system. After the disease has been fought off, some antibodies remain in our bloodstream, protecting us from fresh attacks.

The immune system also involves cell-mediated immunity. This can act alone but normally interacts with antibody immunity to counter the huge range of infectious agents to which animals are exposed in their lives.

In cell-mediated immunity, specialised white blood cells — T-cells — recognise and destroy infected host cells. The presence of a foreign micro-organism in a cell is betrayed by pieces of its molecules on the cell surface. The T-cells recognise the tell-tale signs of infection. They then multiply and become 'activated' killer T-cells. Killer T-cells attach themselves to infected cells and destroy them.

Vaccines

In 1796, Edward Jenner showed that a deliberate cowpox infection made humans immune to smallpox, although he did not know how disease immunity worked. Based on his work, a world-wide vaccination campaign has now probably made the smallpox virus extinct. In 1879, Louis Pasteur accidentally found that weakened cholera bacteria made chickens immune to infection from the normal form. This led to the development of vaccines against many diseases, not just smallpox and cholera. Today, vaccines are used to control a number of life-threatening diseases, including measles, diphtheria, tuberculosis, polio and tetanus. Veterinary vaccines prevent a variety of diseases that endanger livestock and domestic animals. Vaccination mimics an infection. The vaccine triggers the immune system's memory without exposing the host to the risks of infection.

Although many vaccines have been developed, there are still problems. Some vaccines are unsafe or do not work as expected. Some disease agents mutate rapidly, changing to avoid the body's defences. This means that some vaccines are only useful for a short time. Many vaccines are expensive to produce and this is a concern for developing countries, which need them most. Biotechnology is advancing and we are learning more about the body's immune system. These developments will allow cheaper, safer vaccines.

Vaccine production has not changed much since last century. Most vaccines are made of killed micro-organisms, inactivated bacterial toxins or weakened (attenuated) live organisms. All of these have their ability to cause disease reduced. Each vaccine preserves the part of the disease agent that triggers the immune response. The choice between killed or weakened vaccine depends mainly on what type of immune mechanism is normally triggered by a particular disease. Live vaccines tend to cause cell-mediated and secretory antibody immunity. Killed vaccines tend to induce circulating antibody immunity.

Normally it is just a few special molecules on the surface of the infecting agent that trigger the immune response. The agent may be a simple virus that consists of only a few different molecules. It may be a bacterium composed of tens of thousands of different molecules. It may be a parasite with hundreds of thousands of different structural molecules. In all cases, the immune system responds to a few key units.

These key units are termed protective immunogens and it is these that the immune system 'remembers'.

Because the immune system acts only on the few protective immunogens, most of the other parts of killed vaccines are not needed. Indeed, the extra material may be the cause of the bad side effects seen with some vaccines. There is also some risk that the weakening process may be incomplete and that fully active agents will survive.

It would be much safer if vaccines were made of only the protective immunogens. These 'sub-unit' vaccines would have just a few different molecules from the original disease agent. Genetic engineering offers a highly efficient means of producing vaccines made only of these proteins. This will reduce the risks of harmful side effects and the chance of actual infection.

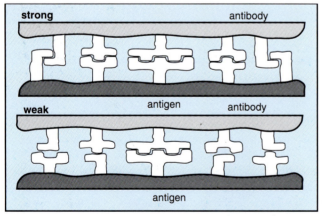

Figure 3.4 Each antibody producing cell makes just one type of antibody that recognises and binds to one particular place on the antigen. The surface of the antibody has sites that match the shape of the antigen molecule. The binding can be weak or strong depending on how well the sites match.

Monoclonal antibodies

As each antibody binds to and attacks only one particular antigen, specific antibodies have many potential uses. As well as protecting against disease, they can also provide diagnostic tools for a wide variety of illnesses. They are able to detect the presence of drugs, viral and bacterial products, and other unusual or abnormal substances.

Scientists have been trying to produce large quantities of antibodies for specific uses for some time. Usually a laboratory animal is injected with an antigen. The animal's immune system produces antibodies which are then collected from the blood serum. There are two problems with this method. Firstly, the blood serum contains other, unwanted substances. Secondly, it produces only a very small amount of useable antibody. Antibodies produced in this way come from a preparation containing many kinds of cells and hence are called polyclonal.

In 1975, Georges Kohler and Cesar Milstein, of the Medical Research Council Laboratory of Molecular Biology in Cambridge, England, carried out some landmark experiments. They used some lateral thinking and hit upon the inspired idea of fusing antibody-producing cells with certain cancer cells to generate cells that only secrete a single kind of antibody.

Cells from a cancer tumour can reproduce endlessly in cell culture. Mammalian cells that naturally produce an antibody were hybridised with cells from a tumour. The result was a 'factory' that worked around the clock. The hybrid animal cells are called 'hybridomas' and continuously produce antibodies. These antibodies are called monoclonal because they come from only one type of cell — the hybridoma cell. Cell culture techniques allow scientists to grow a pure culture of hybridomas. All the cells in the culture will produce the exact same antibody.

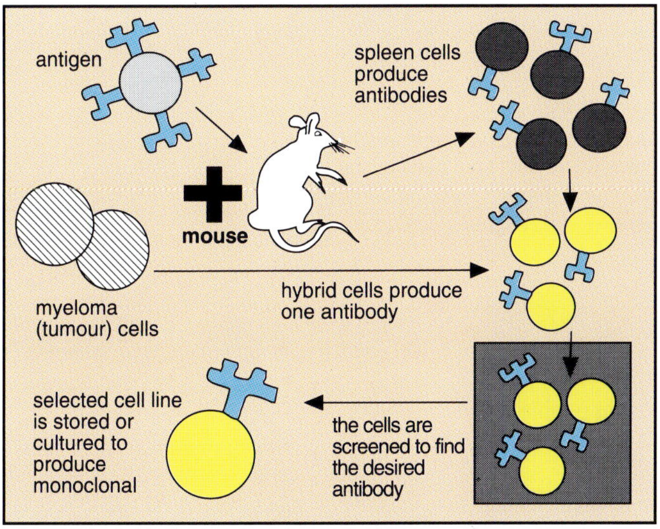

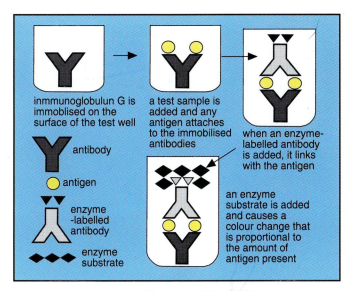

Figure 3.5 Making a monoclonal antibody

Preparing enough monoclonal antibody to the antigen of interest is a first step. A suitable procedure to get good antibody–antigen binding is then needed to develop a diagnostic test for the substance. It is essential to get the components stable and make the method robust so that anyone can obtain results when using the test.

The usual method for small molecules such as pesticides is a 'competitive' ELISA test. ELISA stands for enzyme-linked immunosorbent assay. In an ELISA test, a microwell is precoated with an antibody. The sample to be tested is then added and any antigen attaches to the antibody. A second antibody, labelled with an enzyme, is then added and attaches to the antigen now immobilised on the well. Unbound reactants are then washed away and a colourless enzyme substrate is added. Colour develops as the enzyme acts on the substrate. The colour intensity is measured with a photometer, the more intense the colour, the higher the antigen content in the sample. There are variations in the way ELISA tests are done. The method used will depend on the materials available and what needs to be tested.

Insulin

Insulin is essential in the control of blood sugar levels. Diabetes mellitus is a disease where people do not make any insulin themselves. It kills many people in the developed world each year. Insulin has been used in the treatment of diabetes mellitus since 1922 when Leonard Thompson became the first human to receive an injection.

Figure 3.6 The ELISA procedure

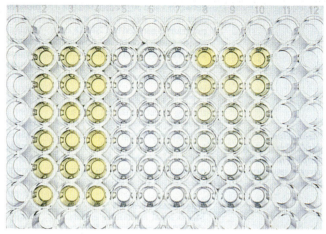

Figure 3.7 The results of an ELISA test. The coloured wells on the left indicate a positive reaction while the clear wells in the centre show a negative reaction. The different shades of colour in the wells on the right indicate different strengths of antigenic response.

There are 250,000 Australians who have been diagnosed as having diabetes and 95,000 of them are insulin dependant. However, there may be up to a quarter of a million people with the disease whose condition has not been diagnosed.

Figure 3.8 Insulin for the treatment of diabetics is now being made synthetically with the help of genetic engineering

Pancreas glands from cattle and pigs were originally used as a source of insulin. Insulin was extracted from the glands and refined to increase the concentration. Improvements in technology then led to the development of single-species and highly purified insulins. These advances mean less side effects like injection-site reactions and insulin resistance. However, animal insulins are different from human insulin. The possibility that diabetics may develop sensitivity to animal insulins could prevent their use. This possibility has led to the search for methods to produce human insulin on a commercial scale.

Human insulin can made chemically but costs too much to be widely used. Biological techniques may hold the answer. Human insulin has been produced commercially by genetic engineering and by biosynthesis from pork insulin.

Biosynthesis involves changing one of the amino acids in pork insulin with enzymes. This results in a purified insulin called human monocomponent insulin. Insulin

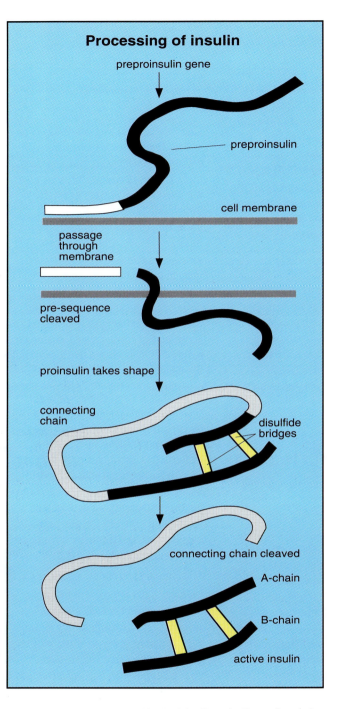

Processing of insulin

preproinsulin gene

preproinsulin

cell membrane

passage through membrane

pre-sequence cleaved

proinsulin takes shape

connecting chain

disulfide bridges

connecting chain cleaved

A-chain

B-chain

active insulin

Figure 3.9 Genetically engineered insulin is produced by working backwards from the amino acid sequence of the two insulin chains and constructing synthetic genes for each chain which can then be introduced into bacteria

produced by this method may still contain very small amounts of pancreatic impurities, such as pork preinsulin, because it is of pork origin.

Human insulin goes through a number of changes before it enters cells from the blood stream. The first protein formed is preproinsulin. This form of protein enables the molecule to move across cell membranes. As it does, the part of the protein that enables it to move through the membrane — you could say the 'pre' part — is lost leaving proinsulin, a storage form of insulin.

The proinsulin molecule contains 84 amino acids arranged in a complicated loop. To convert proinsulin to insulin, enzymes in the pancreas chop off 33 amino acids. This leaves the active human insulin molecule — two chains connected by disulfide bridges. The A-chain of human insulin has 21 amino acids and the B-chain contains 30 amino acids.

It would be inefficient to synthesise insulin by starting with the gene that makes preproinsulin as too much secondary processing takes place. To overcome this problem, scientists have synthesised a DNA molecule that encodes both the A and B chains of insulin. The base sequence (or codon) ATG specifies the amino acid methionine. It was inserted between the two regions of DNA encoding the two insulin chains. This piece of synthetic DNA was then joined to a bacterial gene, again with the ATG codon at the junction.

This constructed gene directs the production of a mixed protein when introduced into the bacteria. The mixed protein consists of the bacterial protein and the two insulin chains, all joined by the amino acid methionine. Neither of the insulin chains contains methionine. Cyanogen bromide destroys methionine and, when it is added to the mixed protein, releases the individual insulin chains from the bacterial protein. The chains are then joined using a reaction that forms disulfide bridges between cysteine amino acids in each chain.

Of all insulin used in treating diabetes, 92 per cent is synthesised from the human insulin gene being expressed in either *Escherichia* (*E. coli*) or yeast. The remaining 8 per cent is extracted from cattle. It is predicted that eventually all insulin in Australia will be obtained from genetically modified organisms.

Human Growth Hormone

If you look around a classroom it is easy to see that people vary in many ways: height, weight, skin colour and eye colour are good examples. Things that vary in this way are often the result of a combination of many factors. An individual's height, for instance, depends on the heights of his or her parents and grandparents, on nutritional factors and on the function of proteins or hormones from different glands such as the pituitary, thyroid and adrenals.

One of the most important of these proteins is Human Growth Hormone (HGH), which is made in the pituitary gland. Abnormalities in growth can occur in several ways. Excess production of HGH can result in an individual being taller than expected

and having serious bone and joint problems as she or he gets older. Individuals who are shorter than expected may have HGH problems due to an underactive pituitary gland.

One way to treat this condition is to collect the pituitary glands from cadavers (dead bodies) and process them for the hormone. To treat just the most severe suffers of HGH deficiency in Australia requires collection of tens of thousands of pituitary glands each year. Research has shown that while the pituitary growth hormone from cadavers promotes growth in hormone-deficient children, they still do not reach a normal height. In one study, about half of those treated reached a height that was within the low end of the height range of the 'normal' population.

In 1981, biologists used genetic engineering to develop *E. coli* bacteria that produce small amounts of HGH. This biosynthetic growth hormone appears to be equivalent to the highest quality natural HGH. Rates of height and weight gain with biosynthetic and pituitary hormones have been virtually identical.

Cadaver pituitary growth hormone is no longer used because a lethal, slow-acting virus infection has now been found to occur in some pituitary glands. In the early 1980s, some young people who had been treated with cadaver pituitary growth hormone became infected with the virus.

Genetic engineering now makes it possible to produce unlimited quantities of the biosynthetic hormone. This process produces a

Figure 3.10 CSL produces human growth hormone more cheaply and reliably by genetic engineering

hormone that gives better results than obtained from cadaver pituitary extracts and eliminates concern about transmission of viruses.

The Commonwealth Serum Laboratory (CSL) in Australia is developing a process that involves using mammalian cells rather than bacteria to make the human growth hormone. The process is chemically simpler and will produce greater quantities than are possible with bacteria. It should lead to Australia producing enough of the hormone for its own use and to sell overseas.

Genetic therapies

'Somatic' gene therapy involves injecting genetically engineered cells into the body to correct the function of defective cells. Scientists are developing genetic substances that can be inhaled to correct faulty cells in the lungs of cystic fibrosis patients. This type of gene therapy is considered to be similar to the use of drugs and poses few moral or ethical dilemmas.

'Germline' gene therapy, however, is a potentially frightening tool involving the introduction of new genes to replace defective genes in newly formed (single-celled) embryos. If successful, the offspring of the resulting individuals would pass on the new trait to their descendants.

For ethical reasons, genetic scientists from around the world voted at the Ciba Foundation Symposium in Berne, Switzerland, in June 1989, to ban any germline research on human embryos. Many scientists consider they do not know enough at this stage to introduce germline therapy. Particularly, the scientists do not know whether a particular genetic manipulation that is programmed to have a good effect in a person might have unexpected, unimagined bad effects. For instance, germline therapy has been used to produce larger pigs, cattle and mice. The offspring often experience devastating side effects in adulthood, such as crippling arthritis and imbalances between protein and fat levels.

The Human Genome Project

Genetic engineering is based on the assumption that we know two things: the sequences of bases on the DNA molecule and the function of a

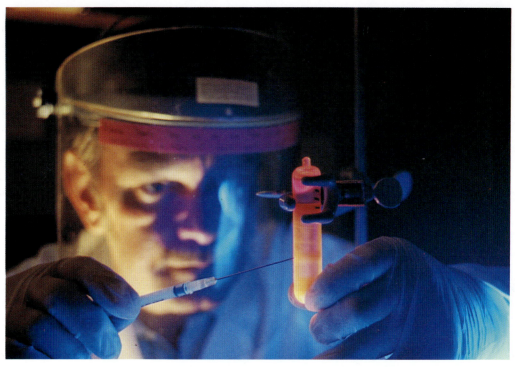

Figure 3.11   A CSIRO scientist separates fragments of DNA

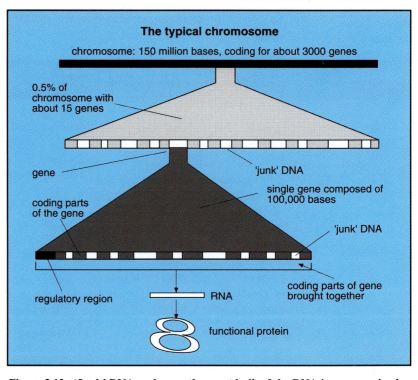

Figure 3.12   'Junk' DNA makes up the great bulk of the DNA in an organism's genome. Production of a functional protein from a gene involves 'editing out' the junk.

particular gene. As our knowledge of these two things increases, genetic engineering will become more powerful. To date, we know the sequences of only a small proportion of bases in only a few organisms and the function of only a small number of genes. In 1977, the sequence of the 5000 bases of a small virus was worked out. By 1990, the sequence of the 250,000 bases of the herpes virus was also known. Only 800,000 of the 4.8 million bases of the bacterium *E. coli* have been determined so far. Human DNA contains about 3 billion bases on each pair of the DNA strands. However, compared to humans, wheat is very complex; it has about 16 billion bases on each DNA strand.

Within the next decade, scientists will gradually break the genetic code responsible for producing every human physical characteristic. In a remarkable scientific endeavour, the Human Genome Project (HGP) will identify all the 50,000 to 100,000 genes that make up the human being: the human genome.

Each of these genes has a specific function within the body. Single genes or combinations of genes cause the production of important chemicals and determine characteristics such as eye or hair colour.

The project, which began in the United States of America in the late 1980s, has identified more than 5000 genes so far. Not all of the genes have been matched to a physical characteristic.

The US Government actually legislated to start the HGP, aiming to determine the ordered sequence of the 3 billion bases of human DNA. Just the cost, without considering the potential effect on the future of the species, makes this is a major undertaking. The initial estimate was about US$12 billion. There will probably be technological advances over the time of the project that will bring the cost down, but the current budget is US$3 billion over the next 15 years. Many laboratories around the world are working on the project.

On top of this effort, the project needs major advances in computing. Today's supercomputers do not have the capacity necessary to process and match the information flowing from the project. The scale of the HGP is about as big as the American Apollo project was to put a person on the moon.

How is DNA mapped?
Mapping of DNA begins by placing the DNA in a solution containing one or more enzymes that break it into fragments. In mammals, the DNA is commonly broken into as many as one million fragments. These fragments are separated from one another according to their length by gel electrophoresis.

Once separated, the fragments can be cloned into bacterial plasmids and reproduced in large numbers. The fragments can be worked on further using the same process again until small enough fragments can be separated and their base sequence determined.

Organisms that are closely related share many DNA sequences. For example, it is estimated that only 150 to 200 different gene sequences separate mice from humans. As few as 100 gene sequences separate cattle from humans. The three billion bases on the human DNA code for about 100,000 genes. Many sequences are repeated over and over. Most of the sequences, about 95 per cent, do not appear to code for any protein.

Possible applications of the HGP

Defective genes are responsible for over 7000 genetic diseases, including cystic fibrosis and haemophilia. At present, 300 genetic diseases can be detected through testing in the unborn foetus but only about 35 of these tests use genetic screening methods. Genetic disorders are of three types. Single gene disorders are inherited. They are caused by a change in the DNA structure of a single gene. Chromosomal defects are the result of a change in chromosomal material during the formation of the embryo. Polygenic diseases result from the interaction of many genes and environmental factors. The usual ways that genetic diseases are detected include:

- **ultrasound examination** late in pregnancy to see if organs are formed properly;
- **chromosome analysis** of foetal tissue, usually between the 12th and 17th weeks of pregnancy for women at risk (because of family history or advanced age);
- **assessment of marker proteins**, usually carried out in Australia on all newborn babies. It determines if certain amino acids that affect metabolism are present in normal quantities; and
- **DNA probes**, a genetic screening technique used to detect major structural damage to genes.

As a result of the HGP, it may be possible one day to identify many more of the genes that are responsible for disease. If so, then it should also be possible, by manipulation, to prevent those genes from causing disease in children whose parents carry the disease-causing gene but do not have the disease (carriers). Most would see this as a positive use of scientific knowledge. Another possibility could be to determine whether individuals have a particular gene sequence that makes them susceptible to certain diseases as they age. At this stage, scientists are hoping information gained from the HGP will allow them to identify people carrying genetic diseases and provide early, more beneficial, treatment for them.

There are potentially negative aspects of the HGP. Once the genes responsible for human characteristics are identified, it will be possible to take a cell from a person and perform tests that will give considerable insight into that person, including their susceptibility to disease, their race and perhaps even their intelligence.

There is really no doubt that many people are going to want access to the results of these genetic tests. For example, insurance companies may use this information to refuse insurance to anyone who has a genetic condition that may be costly to treat. Genetic screening by employers or insurance companies raises privacy and discrimination issues. These issues should be debated by the community now so that governments can make laws that reflect the wishes of the people and ensure that we control our future.

In-vitro fertilisation

The ethical and practical questions regarding the use of new medical technologies have, in general, not been tackled before they have been used. *In-vitro* fertilisation (IVF) is a good example. The technology was introduced before many important issues were addressed and the public's perception is now clouded by the images of grateful, happy mothers cuddling healthy babies. No studies have been done to see if the technique has any negative effects.

In 1978, the first child to have resulted from IVF and embryo transfer (ET) was born. By the late 1980s, about 15,000 babies were born using this new technology. The technology was designed to overcome the problems of infertility in women due to blocked Fallopian tubes. It is now used to overcome infertility in both men and women due to a variety of disorders.

The term *in-vitro* fertilisation simply means fertilisation in glass and refers to the process of bringing together egg and sperm cells in a glass receptacle. Embryo transfer refers to the transfer of the embryo from one place, the test tube, to another, a living body. Amazingly, both IVF and ET were attempted in non-humans in the late nineteenth century. It was not until 1965 that the techniques were applied to humans.

Fertilisation of the human egg takes about 24 hours. In the controlled condition of IVF there is currently about an 80 per cent success rate of fertilisation. The major factors limiting the success of clinical IVF today are the continued growth of the embryo after fertilisation and the losses that occur after the embryo is transferred into the uterus.

Through IVF we have a much better understanding of many aspects of the reproductive process and these have led to the development of new approaches to contraception. At present, contraception relies on two main methods, condoms and the pill in its various forms. IVF research has opened a number of other possibilities. For example, the surface of the sperm is covered with antigens. Anti-sperm antibodies are known to occur naturally in some cases of female infertility. The production of an anti-sperm vaccine may be a new approach to contraception.

There has been an increased understanding of the mechanisms and diagnosis of infertility too. IVF has been a major contributor to understanding and treating male infertility due to a low sperm count. Infertility as a result of weak or few sperm can now be overcome by sperm micro-injection into the egg. Another aspect under study is the ability of the sperm to penetrate the egg or to pass through the mucus lining of the cervix.

Currently the success rate of IVF is about three live births out of every 100 eggs collected. This is considered low so researchers are trying to understand and improve the success rate. From this they are gathering information on the nutritional requirements of human pre-embryos. A pre-embryo is the fertilised egg before it is placed in the woman's uterus, a process called implantation.

Up to 30 per cent of pre-embryos in IVF are chromosomally abnormal. There is no way yet of detecting embryonic abnormalities, especially of a genetic nature, earlier than 8 to 16 weeks into a pregnancy.

Researchers hope their work will allow detection of such abnormalities in the first few days, before implantation.

There are several moral questions to consider with IVF. Firstly, in a grossly overpopulated world, is it sound practice to spend the limited resources on developing new ways to produce more humans? Secondly, the scientific research involved in IVF and ET involves the destruction of human embryos. When does the embryo acquire a moral status that prevents its purposeful destruction? When does an individual begin to exist? And what should happen to stored frozen embryos? In 1992, there were more than 9000 frozen embryos stored in IVF units in Australia and New Zealand. It seems that few parents want the embryos destroyed or used for research.

Thirdly, the sperm and egg used to produce an embryo do not necessarily originate from the future 'parents' of the child that results. Who are the legal parents — the genetic parents or the birth parents? Should genetic parents and child know each other's identity? Few couples in NSW are prepared to risk donating their embryos to other infertile people because under existing state law the egg donor remains the legal mother.

Who should answer these questions? Should the research and IVF programs continue without these questions being resolved? IVF programs also involve experimentation. Who is the subject of the experiment? If the subject is the unborn, is it moral, or legal? Is the effect on the adult participants being taken into account?

DNA fingerprinting

With the exception of genetically identical twins, the chemical structure of DNA differs in every person. In 1984, Professor Alec Jeffreys at Leicester University discovered that a number of sequences of genetic information exist within the human DNA molecule which vary greatly within unrelated individuals. These sequences are highly repetitive and can be found many times along the chain-like structure of DNA. The lengths of each repeated sequence, the number of repeats and their exact location within the molecule are quite characteristic of an individual. The sequences can be 'visualised' by a series of laboratory techniques including electrophoresis, construction of a radioactive DNA probe and exposure on an X-ray film.

DNA fingerprinting is the name given to a scientific testing process that positively identifies individuals from their genetic material. It is also known as genetic fingerprinting or DNA profiling. The technique was first used in 1987 in Britain to track down the killer of two teenage girls.

Information from DNA fingerprinting is admissible in Australian courts. A person may voluntarily provide a sample for testing or a sample may be taken under a court directive if there are already reasonable grounds from other evidence to show that such an invasion of civil liberty is warranted. If the person is not convicted, there are strict regulations about the destruction of samples and information that might identify the person.

Waste that can be used

Whey is a by-product of cheese manufacture. For years it has been seen as a waste product. The main proteins in whey are b-lactoglobulin (45 per cent), a-lactalbumin (20 per cent) and, in cheese whey, the casein-derived peptide (20 per cent). Minor protein components include serum albumin, lactoferrin, immunoglobulins, phospholipoproteins and enzymes.

One area of research involves a mixture of naturally occurring growth factors found in whey. The Cooperative Research Centre for Tissue Growth and Repair in Adelaide and its commercial partner, GroPep, have patented a process for extracting and purifying these.

Growth factors are used in the pharmaceutical and biotechnology industries, where animal cells are grown in culture using growth factors currently derived from foetal calf serum. The potential for whey-derived growth factors in these industries is estimated at between $200 million and $300 million a year.

Australian pharmaceuticals

White blood cells are required by the body to help fight infection. Some years ago an Australian team at the Walter and Eliza Hall Institute of Medical Research, led by Professor Don Metcalf, discovered that production of these blood cells is stimulated by the hormones G-CSF (Granulocyte colony stimulating factor) and GM-CSF

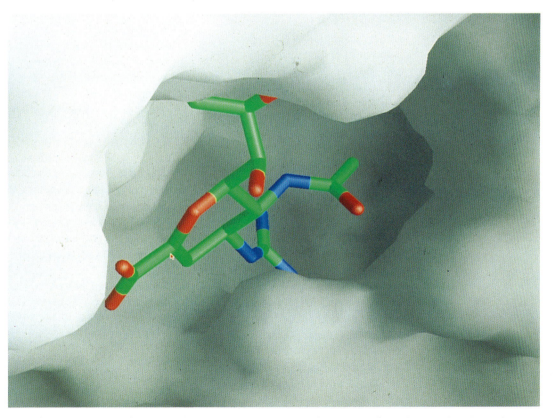

Figure 3.13 A biopharmaceutical which acts by blocking the site where viruses can attach

(Granulocyte-macrophage colony stimulating factor). The former stimulates white blood cell production in cancer patients after chemotherapy or bone marrow transplantation. The latter is similar to G-CSF, but with a broader action.

The hormones are now made synthetically and marketed internationally as drugs. They are revolutionising the treatment of cancer patients and have world-wide sales in excess of US$500 million. Unfortunately, Australia received little financial benefit from this discovery. The hormones were developed by pharmaceutical companies in Europe, the USA and Japan. There has been no Australian participation or ownership.

Professor Metcalf has now made another discovery and Australia is in control of its commercialisation. Ten years after his first discovery, we now have an internationally competitive Australian biopharmaceutical industry. The latest discovery is an important blood cell growth factor, known as LIF (Leukaemia inhibitory factor). LIF stimulates platelet production in cancer patients after chemotherapy or bone marrow transplantation.

The Melbourne-based pharmaceutical company, AMRAD Corporation Ltd, sponsored the research that led to the identification of the LIF hormone. The next phase, the clinical development of LIF, is beyond the financial resources of AMRAD so it has formed an alliance with Sandoz Pharma of Switzerland and Chugai Pharmaceuticals of Japan. AMRAD retains ownership and control over LIF's development.

Australia can expect considerable financial returns. We will also learn about developing drugs to a commercial product. This will have benefits for our future.

1. Find the following words in this chapter. What does each word mean?
   - Genome
   - Polyclonal
   - Somatic
   - *In-vitro* fertilisation
   - DNA fingerprinting

2. Each paragraph in a well written passage will contain a single sentence which gives its main idea. Turn back to the section of the chapter which describes monoclonal antibodies. Write down the key sentence in each paragraph. Look at the other sentences in the paragraph. They will develop or support the main idea given in the key sentence. Use the ideas from the key and support sentences to write a point form summary of this section. Remember, a point form summary separates main and supporting ideas.

3. Use the subheadings within this chapter to show the structure of its ideas. Use the key sentences from each paragraph to summarise the chapter within this structure.

4. Explain how vaccination works.

5. Describe the two ways in which the immune system protects us from disease.

6. Genetically engineered vaccines have a number of advantages over the kind of vaccine prepared by Jenner in 1796. List three of them.

7. What is the importance of the work of Kohler and Milstein?

8. Describe diabetes and its treatment.

9. Describe the scope and importance of the Human Genome Project.

10. Why have scientists stopped germline gene therapy on human embryos?

11. Comment on some of the dilemmas surrounding *in-vitro* fertilisation and embryo transfer.

12. Paragraphs are often tied together by 'pointer words'. Pointer words are words which refer to other parts of the paragraph. They are small words such as 'he', 'she', 'it', 'they', 'this', 'each', 'who' or 'that'. Turn back to the section on the immune defence system. Take a pencil and put a circle around each pointer word. Draw an arrow from the circle to the words to which the pointer refers.

13. 'Connectives' are words that link ideas. Connectives are joining words which show the relationship between the ideas they connect. Choose words from the following list to join the pairs of sentences below

- similarly
- after
- where
- and
- furthermore
- eventually
- meanwhile
- but
- although
- often
- because
- as
- which
- or

Write the joined sentences. You may need to change some other words to combine the sentences. You may use the same word more than once and you can use more than one word from the list in each pair of sentences. The first pair is done for you

Diabetes mellitus is a disease
People with diabetes cannot make their own insulin.
Diabetes mellitus is a disease <u>where</u> people cannot make their own insulin.

Insulin is essential in the control of blood sugar levels.
Diabetes sufferers cannot control the level of sugar in their blood.

Diabetes kills many people in the developed world every year.
Sufferers can be helped by injections of insulin.

Leonard Thompson was the first person to have an insulin injection.
He was injected in 1922.

Two hundred and fifty thousand Australians have diabetes.
Ninety-five thousand of them are insulin dependant.

Another 250,000 people may have the disease.
These people have not been diagnosed.

Chapter 4
Food

Food is an essential part of everyone's life and the earliest uses of biotechnology were in food production. Micro-organisms produced the yoghurt, beer, wine, sauerkraut and cheese in your refrigerators as well as the olives, bread, soy sauce and vinegar in kitchen cupboards.

Biotechnology techniques involving fermentation have produced food since ancient times. Wine and beer-making probably used yeasts before 6000 BC, yeasts aided in producing leavened bread around 4000 BC, and the Aztecs were harvesting algae from lakes as a source of food before AD 1500.

The earliest known evidence that ancient people drank beer comes from the discovery of a double-handled jar

Figure 4.1 Many common foods have been produced by biotechnology

more than 5000 years old at Godlin Tepe in the Zagros Mountains, western Iran. Tests of pale yellow deposits found in unusual criss-cross grooves on the inside of a portion of the jar suggest they are deposits of calcium oxalate, a substance that settles out when barley beer is stored or fermented. The finding supports other evidence of beer. Sumerians had lived at the site and at several nearby sites in Iraq where similar jars have been found. The Sumerian symbol for beer shows a jar with groove-like markings. Researchers had already found that barley was common at the site and believe the grooves were put in the jars to remove the bitter-tasting calcium oxalate from barley beer.

When the Sumerians brewed their barley beer, they used a biological process. They combined yeast cells and nutrients (barley) forming a fermentation system in which the yeast consumed the nutrients for their own growth and produced by-products (alcohol and carbon dioxide gas). Although more sophisticated, the same principle of combining living matter (whole organisms or enzymes) with nutrients under the conditions necessary to make the desired end product forms today's biotechnology.

The fermentation process in early food production was done without the knowledge that micro-organisms were involved. The micro-organisms came as a contamination from the air or ingredients used. Later, previously contaminated 'starter' material was used, but still without the knowledge that micro-organisms were present, or their function. All the early fermentation was the result of activity of a mixed culture of cells.

About the mid-1800s, Louis Pasteur and others discovered that micro-organisms were responsible for alcohol and vinegar production in fermentation. Pasteur was the first to scientifically study fermentation. From his studies of fermentation in wine and milk, he found that fermentation was due to the activities of micro-organisms. He found that the micro-organisms that produced 'good' fermentation were different to those that produced 'bad' fermentation. Pasteur also found that by heating wine to a temperature of 50 to 60°C and holding it at that temperature for a few minutes, the wine did not go off, and retained its flavour. This process became known as pasteurisation and, although it is no longer used for wine, it is still used in the treatment of milk.

Pasteurisation was widely used by the late 1800s. The aim was to reduce the level of contamination of unwanted micro-organisms in food processes. In 1896, the first pure culture of micro-organisms produced beer at the Carlsberg brewery in Copenhagen.

Better bacteria make better cheese

Cheese-making is an old process and probably started thousands of years ago as a way of preserving the nutritional value of milk. Today, fermented dairy products account for 20 per cent of the total value of fermented foods around the world, and cheese is the largest component.

To make cheese, milk is inoculated with lactic acid bacteria, strains of *Lactococcus lactis*. The bacteria increase the acidity of the milk, which protects it from infection by unwanted or harmful organisms. An enzyme found in the guts of calves, rennin, is also added to the milk, causing it to coagulate. Solid particles (the curd) are separated from the whey, or watery liquid. The curd is then cooked, pressed into shape, salted and allowed to ripen. During ripening, the fats, protein and other compounds are hydrolysed. This gives the distinctive cheese flavour. Micro-organisms have been used for a long time to make cheese. However, modern methods are improving quality and quantity. An important change in modern cheese-making is the addition of microbial enzymes to accelerate ripening.

Australia produces about 190,000 tonnes of cheese each year, valued at $380 million. On a commercial scale, the process involves juggling the blend of living organisms within the vats that trigger the conditions needed for curd formation. Problems occur with the bacteria used. They cannot be controlled with the same certainty as simple chemical reactions because they are living organisms.

There are three problems with bacterial cultures: they are living cells that compete with other species, they show genetic variability, and they are vulnerable to parasitic

Figure 4.2   Cheese-making is a major industry in Australia and depends on the efficiency of the bacterial cultures used to ferment milk sugar (lactose) into lactic acid

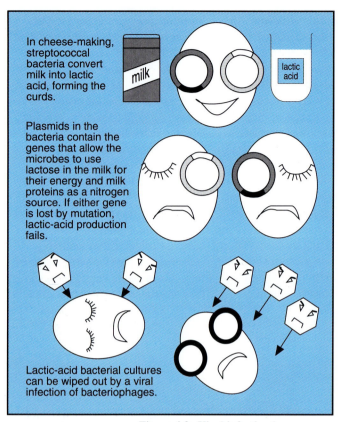

In cheese-making, streptococcal bacteria convert milk into lactic acid, forming the curds.

Plasmids in the bacteria contain the genes that allow the microbes to use lactose in the milk for their energy and milk proteins as a nitrogen source. If either gene is lost by mutation, lactic-acid production fails.

Lactic-acid bacterial cultures can be wiped out by a viral infection of bacteriophages.

Figure 4.3 Viral infection by bacteriophage in cheese-making

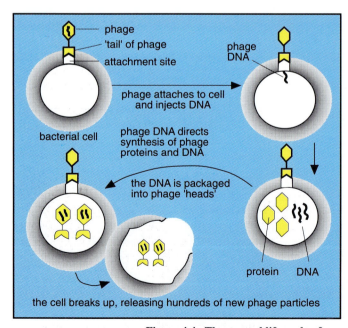

phage
'tail' of phage
attachment site
phage DNA

phage attaches to cell and injects DNA

bacterial cell

phage DNA directs synthesis of phage proteins and DNA

the DNA is packaged into phage 'heads'

protein   DNA

the cell breaks up, releasing hundreds of new phage particles

Figure 4.4 The normal life cycle of a bacteriophage

attack, particularly from viruses. A team of food technologists from CSIRO's Division of Food Science and Technology has been working on these problems.

Bacteriophage resistance

Viruses that attack bacteria are called bacteriophages or phages for short. They inject their own DNA into the bacteria and so force the bacteria to make many copies of the virus. The bacterial cells then split open to release the new viruses and the infection spreads. When *Lactococcus* bacteria are infected, their production of lactic acid decreases.

In the 1950s, rotating starter cultures solved the problem. In this way, the starter culture in use always had a low level of bacteriophage contamination. However, as the volume of cheese being produced increased, this practice became impractical. Researchers at CSIRO came up with an improvement. They exposed bacterial cultures to whey from curd formation. Any starter strain that can reproduce in the presence of whey must be relatively phage-resistant because whey contains bacteriophages. This method is still in use in most large Australian cheese factories. Unfortunately, the method still leaves a level of uncertainty and so the research for a more reliable solution continues.

The CSIRO team has used genetic engineering to improve phage resistance. Bacteria resist phage attack in at least three ways. The first involves a mutation of the bacterial cells; the outside of the cell can change so that the phage cannot attach itself and hence cannot inject its DNA into the host and cause infection. The second is used if a phage does succeed in attaching. The bacterium makes enzymes, restriction endonucleases, that recognise DNA from invading cells and 'cut' it. Some phages make their DNA mimic the host to overcome this attack. In the third, the host cell dies, but without splitting. In this case, the virus can multiply but cannot escape the cell to infect other bacteria.

Strain identification

Gel electrophoresis sorts different strains of bacteria. Scientists cut bacterial DNA into 20 to 30 large fragments using restriction endonucleases and then separate them according to size using gel electrophoresis. The unique pattern each starter strain leaves on the gel has helped identify about 50 different strains. Those with phage resistance still have the pattern of their original strains. This quick and efficient process enables scientists to check that they still have the same fermentation and flavour-producing properties as the parents.

Lactic acid production

When the starter bacteria convert lactose to lactic acid, the bacteria produce a number of intermediate compounds. These compounds accelerate the action of certain enzymes which in turn increase the rate of acid production. The research team is currently trying to work out which genes code for the enzymes that control the making of lactic acid. So far, four genes have been identified but it is expected that another eight or so will be found.

SIROCURD — another way of making cheese

When we say cheese, we usually have in mind a piece of cheddar. Each year the world produces about three million tonnes of cheddar cheese. Australia's contribution is about 200,000 tonnes.

Most of this cheese is produced using the batch method that cheese makers have used for centuries. This carries the risk of variation between batches.

As discussed earlier in this chapter, the traditional process begins when the cheese maker fills a large vat with milk and throws in a starter culture of lactic acid bacteria. These

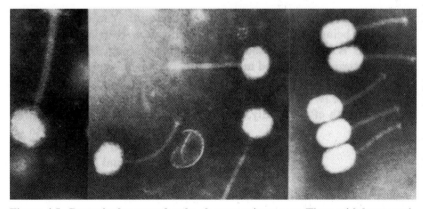

Figure 4.5 Bacteriophages under the electron microscope. Those with hexagonal heads appear to create bigger problems for cheese makers than the ones with elongated heads.

bacteria grow on lactose to make lactic acid. The next stage involves the addition of rennet. This enzyme destabilises casein to form a gel. The lactic acid bacteria drives water (whey) out of the curd.

In making hard cheese like cheddar, the curd is cut into small pieces, heated and stirred. The task of the cheddar cheese maker is to get enough moisture out of the curd to make a firm, dry cheese. Ultrafiltration is used at this stage. It involves the use of pressure to push liquids through membranes having microscopic pores.

In cheese-making, ultrafiltration is used to concentrate fat and protein from pasteurised milk. A number of soft cheeses are made using this method. Examples are fetta and camembert. Some semi-hard cheeses, such as mozzarella and blue vein, also use this method. Ultrafiltration does not work as well with hard cheeses.

CSIRO's Division of Food Science and Technology helped develop another method that may offer the advantage of continuous production, consistent quality, increased yield and export earnings from the sale of equipment. The automated, continuous method of making cheddar cheese is called APV-SIROCURD.

It was known that four factors determine the quality of cheddar cheese: the composition, the acid production, the curd structure before salting, and the maturation temperature.

Research by CSIRO found that the level of lactose acid left in the cheese after ultrafiltration, along with the physical damage of the milk components caused by pumping, were responsible for the loss of both structure and flavour in cheddar cheese made by the traditional method.

The SIROCURD process is a modification of ultrafiltration. It begins with the milk being concentrated and the level of lactose being reduced. During this part of the process, the equipment has been modified to reduce physical damage and keep the milk moving rapidly.

The next step is to add a cheese starter culture to increase acidity. This is done by taking 10 per cent of the milk, heat treating it to kill all the bacteria and then adding the lactic acid bacteria. This part is held for a few hours to allow the acidity to increase. It is then added to the rest of the milk. Because this is a continual process, this bacteria-inoculated milk is added to the milk coming through later.

Rennet is added and the cheese gel is cut 16 minutes later. This gel is continually being discharged from the machine for cutting. The time between the adding of the rennet and the cutting is critical to the final curd composition, cheese yield and cheese quality.

The cooking stage is also critical. In the SIROCURD process the gel cooks in slowly rotating drums at 35 to 40°C until the curd is suitable for the hard cheese. The final stage is the separation of the curd from the whey.

The SIROCURD process takes about 100 minutes for the milk to reach the stage of a curd suitable for cheddaring. The conventional process takes twice as long.

The SIROCURD process enables cheese to be made on a continuous production line. Another claim is that up to 8 per cent more cheese can be produced from the same amount of milk. The world's first such plant was built in Victoria and then the process was adopted overseas. A factory in the USA applied the technology and converted about 48,000 litres per hour, or 800,000 litres of milk a day, into cheddar cheese using the SIROCURD system.

Milk for the lactose intolerant

Lactose is a form of sugar in milk. Many people are intolerant to lactose — about 10 per cent of Australians cannot digest lactose easily because their intestines produce little lactase, the enzyme that breaks down lactose. Instead, bacteria in the gut ferment the lactose. The gas that forms as a result can cause discomfort, distension of the abdomen, flatulence, vomiting and diarrhoea. People who are intolerant do produce sufficient lactase enzymes as babies, but lose this ability shortly after infancy.

Lactose intolerance is inherited. It affects mainly Asians and Australian Aborigines. A fungus, *Aspergillus oryzae*, is now used to make lactase. The enzyme is purified and chemically bonded onto particles of ion-exchange resin. Passing milk through this resin causes the bulk of the lactose in the milk to split, or hydrolyse, into the simple sugars glucose and galactose. This is the same process that happens normally in the stomach. A group of researchers working with CSIRO's Division of Food Science and Technology has developed the technique in Japan. Although the action of lactase had been known for years, the breakthrough came with the development of the technique for fixing it on resin particles. On the resin, the enzyme is not consumed and can be re-used for months. A milk

Figure 4.6 Equipment for splitting lactose.

factory at Drouin, Victoria, is the first in the world to apply this technology commercially. The factory is making a lactose-free milk powder for the Asian market. Other products, such as whole milk, yoghurt and ice cream, are being investigated.

Browning in fruits and vegetables

Browning is a major post-harvest problem which can result in the downgrading of fresh produce for both export and domestic markets. Work is in progress at the University of Adelaide to decrease browning during processing and engineer improved varieties with low browning capacity.

The biochemical basis of browning in grapes involves polyphenol oxidase (PPO). This has been identified by using a sultana mutant, Bruce's Sport, that does not brown easily. PPO occurs in two forms — active and inactive. In Bruce's Sport, the normal conversion of PPO to its active form is interrupted. PPO genes have now been cloned for grapes, apples and potatoes. Other crops will benefit from the low browning character so gene transfer research is continuing.

Ripe tomatoes

Tomatoes have had a lot of 'improvement' work done on them. Through breeding programs, they have been made bigger, redder, resistant to some diseases and stronger to cope with transport. But they have lost their taste.

Through genetic engineering, Calgene Fresh, a division of Calgene Inc., has developed the Flavr Savr tomato. This tomato remains firm and sweet for seven to ten days longer than ordinary tomatoes without refrigeration or the use of ethylene gas. This has been achieved through the introduction of a gene that neutralises a pectin degrading enzyme. With the new gene, the fruit can be picked later. This means that the ripening and flavour-producing processes occur on the vine.

CSIRO's Sydney laboratory of the Division of Horticulture has isolated genes that lead to softening in tomatoes. The scientists also know which proteins the genes encode. The proteins are cell wall modifying enzymes that become active as ripening progresses. The scientists have isolated the genes, changed them and reinserted them into tomato plants in attempts to alter the pattern of fruit softening. If the timing of tomato ripening can be controlled, we will be able to buy better quality tomatoes in the shops.

Is genetically engineered food different?

Sometime soon you are sure to eat something that has been genetically engineered. Except for the small batch of 'superpigs' slaughtered in Adelaide at the beginning of the 1990s, transgenic plants and animals have not, at the time of writing, reached the marketplace in Australia. All over the country, however, genetically altered plants are coming out of laboratories where they have been nurtured for the past decade and planted in small, contained trial plots on experimental farms.

With a growing number of people in the world to feed, some scientists argue that genetic engineering techniques will enable a solution to the food crisis. The 'greening' of consumers means they now want foods that are free of pesticides and produced in ways that do not damage the environment. All of that points to genetic engineering — but only if consumers will buy it. So far, there have been few protests in Australia against the hundreds of genetic engineering experiments going on around the country. It will be interesting to see what will happen when the first products hit the supermarket shelves.

In late 1992, CSIRO scientists 'challenged' the transgenic potato plants they had been growing at Gatton, about 80 kilometres from Brisbane. In the test, the plants were exposed to a potato virus they were engineered to resist. If these and

Figure 4.7   Genetically engineered potatoes do not look or taste any different to the ones we see in the market

future tests work, the final step will probably be the supermarket shelf. In this case, since CSIRO's commercial partner in the potato experiment is Coca-Cola/Amatil, the most likely place you will find the genetically engineered potatoes will be in a packet of potato chips.

In 1993, a genetically engineered tomato, the Flavr Savr, with delayed ripening ability allowing it to be left longer on the vine, went on sale in the US following its successful testing there and in Britain. Unifoods is now testing the modified tomato in Sydney. The Australian trial aims to produce better tasting tomatoes to ship from Victoria to South East Asia without them going bad on the way.

Although almost no-one argues that humans are at risk from eating transgenic food, there are some scientists who are against the introduction of genetically engineered foods at this stage. They argue that genetic engineering does not always result in predictable modifications to organisms and that the interaction between genes can be highly complex. For instance, insecticidal toxins made by non-food plants introduced into crops through gene splicing are new food ingredients. Their safety should be tested. It could even be possible that breeding plants for resistance to pests may make the plants produce substances that are more toxic to humans than the synthetic pesticides they replace.

An interesting dilemma for some consumers is whether people eating a plant containing an animal-derived gene can still call themselves vegetarian. Scientists in the USA

have successfully used anti-freeze genes from flounder to genetically engineer frost-resistant turnips and potatoes. In Australia, CSIRO researchers at the Division of Tropical Crops and Pastures are investigating the possibility of including such genes in sugarcane. In future, sugarcane could grow in cooler areas of southern Queensland and northern New South Wales. Would using sugar from this cane be the same as eating fish?

In many cases it will be hard to police whether any genetically modified organisms were used to develop a product. Even so, there should be evaluation of all new food products reaching our shelves. With so many products being sold on a world market, there is obviously a need to blend with international approaches.

The Australian House of Representatives Standing Committee report, 'Genetic Manipulation: The Threat or the Glory', did not make recommendations about labelling. The Committee suggested a system in which labelling should be decided case-by-case. The argument is that if a food made through genetic engineering technology is found to be identical to that food made in traditional ways, then it should require no further scrutiny. It will probably fall to the regulatory authorities — like the National Food Authority — to decide whether the food is different and what the consequences are for the consumer.

1. Find the following words in this chapter. What does each word mean?
   - Fermentation
   - Micro-organism

- Pasteurisation
- Inoculate
- Gel

2. Each paragraph in a well written passage will contain a single sentence which gives its main idea. The rest of the paragraph develops or supports the idea expressed in this key sentence. Write down the key sentence from each paragraph in this chapter. The list of key sentences is a crude summary of the chapter.

3. 'Biotechnology is one of the world's oldest technologies.' Give evidence to support this statement.

4. How did pasteurisation change biotechnology?

5. List in point form three arguments for and against the use of genetic engineering for food production.

6. List three problems which make bacterial cultures more difficult to use as part of food production. Comment on ways these problems can be solved.

7. Use a flow chart to show how cheese is produced.

8. What are bacteriophages? Describe a method for dealing with bacteriophages in cheese-making.

9. What is 'lactose intolerance' and how can it be treated?

10. Compare natural and genetically engineered food.

11. List the food you ate yesterday. Show which food involves biotechnology and which food does not.

12. Draw a concept map for this chapter. A concept map is a diagram which shows the main ideas in circles and connecting ideas on lines between them.

13. A precis is a summary written in sentences rather than points. Use the concept map you have drawn to write a precis of this chapter.

14. 'Connectives' are words that link ideas. Connectives are joining words that show the relationship between the ideas they connect. Turn to the section on the Sumerians brewing beer. Using a pencil, draw a circle around each connective in the two paragraphs. Draw a table which shows the connective and how it has been used. The first connective is done for you below.

Connective	Use in this sentence
or	Shows that storing and fermenting barley beer will both produce calcium oxide.

Chapter 5
Animals

Over time, we have tried to improve the productivity, disease resistance, litter size and fertility of domestic animals by selective breeding. Through this technique, animals with desirable characteristics are mated and only the best offspring are selected for further breeding. The main limitation of this procedure is that genes can only be exchanged between animals that can mate with one another and produce viable offspring. Genetic engineering has changed this as it allows the transfer of genes between species.

Transforming animals

Animals are easier to genetically engineer than plants. The embryos of many species can be surgically extracted, treated, and then re-implanted in the mother.

Figure 5.1 Conventional breeding has been very successful in adapting organisms for domestication. Biotechnology uses much more precise tools to improve our breeds.

In the early stages of fertilisation, just following penetration of the egg's cell wall by the sperm, the male and female DNA form separate nuclei, called pronuclei, that slowly move together and fuse. The two pronuclei are visible, and scientists have developed micro-injection techniques, delicately manoeuvring a syringe, to introduce foreign DNA into one of the pronuclei.

The cell is only 150 microns across (1 micron equals one millionth of a metre) — ten cells would span the head of a pin — and the target nucleus is only about 20 to 30 microns wide. When the target is finally hit, about one picolitre (a billionth of a litre) is injected into the pronucleus.

Big mice

Early experiments in the creation of transgenic animals were performed with laboratory strains of mice. These experiments showed that the introduced genetic material became part of the chromosomes of the transgenic animals. The introduced genetic material then passed on to each future generation in the ratios expected for normal inheritance. Most importantly, the new genes were functional in the transgenic mice and their offspring.

One of the classic experiments was done in 1982 by scientists from the Universities of Washington and Pennsylvania. It involved a transgenic mouse containing a human gene encoding the information for growth hormone. A 'promoter' sequence had been fused to the gene which caused the gene to function in response to the stimulus of a heavy metal like zinc.

When this mouse was given low levels of zinc in its drinking water, the gene was activated and resulted in the production of large amounts of growth hormone. Since growth hormone regulates the rate of growth of an animal, this mouse grew much faster. By the time it had stopped growing, it was nearly twice as large as a normal mouse.

An isolated gene altered the physical characteristics of a mouse so

Figure 5.2 A very fine glass needle (at bottom) is used to inject genes into an animal embryo. The egg is held in place by suction from the tube at the top.

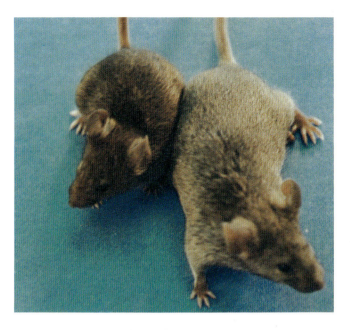

Figure 5.3 American researchers inserted a rat growth hormone gene into mice. The mouse on the right is larger because it has extra growth hormone circulating in its blood.

scientists tried the same technique to change some of the characteristics of farm animals to increase their productivity.

Growing pigs

Studies have shown that the pig growth hormone gene can be inserted into the pig's genetic material, together with a promoter switch similar to that used in the mouse experiment. Unfortunately, the promoter switch is not suitable for field trials because it is activated by dietary copper, which is common in a pig's diet. If a suitable promoter could be found, ideally a harmless chemical not found in a pig's diet, this would allow the producer to switch the gene on and off at appropriate stages in the production cycle and result in leaner pork, produced efficiently.

Pigs have become feral animals in Australia, causing serious land degradation, destruction of crops and predation. There are fears that the release of genetically modified pigs may increase the environmental problem. The key to avoiding this problem for pigs with extra growth hormone lies with the promoter. If an acceptable switch, such as a dietary chemical not occurring in nature, can be used to activate the gene, scientists believe that genetically modified pigs should prove no greater threat to the environment than existing domestic pigs.

Testing for Pompe's disease of cattle

Pompe's is a rare disease of cattle. Calves affected show muscle wasting and usually die from cardio-respiratory failure within six months of weaning. About one in 150,000 human babies are also born with the disease and die within the first year.

For cattle in Australia, Pompe's disease is only considered important in the Brahman breed although it has been found in Shorthorn. This is because the Brahman breed is new to the country. The herd has been built around a very small population, some of which were carriers of the disease.

Those that suffer from Pompe's disease lack a crucial enzyme called α-glucosidase, which is needed to process glycogen. Glycogen contains glucose, which is needed for energy and can only be released by the action of the enzyme.

The disease is inherited. If two carriers mate, one in four will have

Figure 5.4 Calf with Pompe's disease

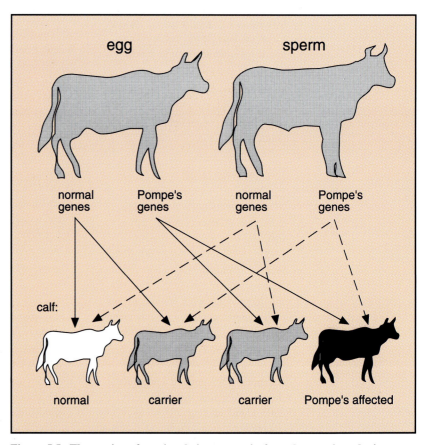

Figure 5.5 The mating of carriers brings a one-in-four chance of producing a calf with Pompe's disease

the disease, two will be carriers and the fourth will be free of the genetic character. It is estimated that about 15 per cent in a randomly mated herd are carriers. Until recently, tests for determining which animals are carrying the disease have lacked precision.

In response to the industry's need for a quick and reliable test, a team of scientists from CSIRO's Division of Tropical Animal Production in Queensland has been working on a genetic test.

In this research, the DNA from normal animals and that from Pompe's carriers was cut using a variety of restriction enzymes and then run through an electrophoresis gel. One particular enzyme, *Msp*I, cuts the DNA in such a way that the DNA from normal animals can be distinguished from Pompe's carriers.

The gene that codes for α-glucosidase differs in the two DNA fragments. The sequence of bases was determined by examining the fragments. Once the sequence was known it was possible to find a specific chemical that could identify the part of the gene that was affected.

The test only needs one microlitre of blood and can be done in about six hours. It can be semi-automated so that large numbers of samples can be tested in a day. The researchers now know that the mutation that has happened to the gene is in one single base. It causes the amino acid glutamine to replace proline, normally found in the protein.

Breeding cattle for the tropics

In tropical and subtropical climates, the Brahman cattle breed, *Bos indicus*, is better adapted than the European cattle,

Bos taurus. The Brahman is resistant to problems like heat, worms, ticks and diseases. It does not seem to be affected by changes in the quality and quantity of food typical of these climates.

The Brahman does have some drawbacks though. When conditions improve, it is not able to grow and reproduce as fast as other breeds. For some years, CSIRO's Division of Tropical Animal Production has been working to breed cattle with the best of both species.

Cattle growers generally breed with the bull that grows fastest in the environment for which selection is being made. The CSIRO work has shown that a better selection can be made using animals with higher fertility, as well as growth. After a long search of breeding records at CSIRO's research station near Townsville, it appears that the cross between Brahman and European cattle produces good growth and high resistance to the tropical environment.

In practice it is not feasible to run only Brahman-cross cattle because there are not enough pure-bred cows. This shortage led to the realisation that an even greater improvement can be made by using a third cattle type to cross with the Brahman cross. To get a third type, scientists went to Africa to look at cattle not yet available in Australia. The African cattle breeds are highly productive, mainly because of their high reproductive rates. Two types, the Boran and the Tuli, were chosen for introducing into Australia.

KAKADU
Wetlands at Risk

WILDERNESS SOCIETY AUSTRALIA

Plumed Egret, Kakadu National Park, Northern Territory, World Heritage Area. The vast wetlands of Kakadu support more than 250 species of birds, 10% of which are rare or endangered. The South Alligator River is the last remaining major river system in Kakadu that is relatively undisturbed, but exploration and proposed mining operations now threaten to pollute it. Help save Kakadu's wildlife and the South Alligator River by writing to Senator Graham Richardson, Minister for Environment, Parliament House, Canberra, voicing your concern. For further information contact The Wilderness Society: Sydney, (02) 267 7929; Melbourne, (03) 670 5229; Hobart, (002) 34 9366; Canberra, (062) 49 8011; Adelaide, (08) 231 6586; Brisbane, (07) 229 4533. Photograph by Tom and Pam Gardner.

Figure 5.6   We need to conserve genetic variety in all animal life

A new quarantine station on the Cocos Islands allows the importing of such cattle. In 1989, the first Boran and Tuli embryos were transported to the Cocos and implanted in cows from Western Australia. After six months, the calves were imported to Australia where they were further bred. In 1993, the first embryos from these breeds were sold to graziers. The outcome should see an improvement in the productivity of our tropical and sub-tropical cattle herds.

Conserving animal genetic resources

We recognise the need to conserve wildlife. It is also important to conserve the variety of genetic resources of our domestic animals and their relations. These animals are the result of breeding along specific lines for as many as 100,000 years.

As the animals have been bred, the environment in which they are kept has changed. For example, we now have irrigation and improved pastures on which to graze sheep and cattle. As the environment has changed, the animals have been bred to adapt to the improved environment.

The process of improving these animals is one in which a number of so-called useless genes have been bred out. If our needs change or the environment changes, it may be difficult to get these animals to prosper. The larger the gene pool from which we can choose animal characteristics, the greater will be our chances of breeding new stock for any changes. Already the ancestral wild species of much of our present-day stock have become extinct and other species are seriously lacking in genetic diversity.

Genetic variation can be lost in two ways. Some breeds are lost through dilution or replacement of their genetic material. Some breeds are lost because we concentrate on individual animals during processes like artificial insemination.

The first step in preserving animal diversity is to find out what genetic resources are available, where they are and how they are adapted to their particular environment. Once these are identified they need to be documented, evaluated and maintained. A major problem with conserving breeds is that many of the older breeds are no longer productive enough and therefore there is economic pressure on farmers to replace them.

It is possible to create gene banks for long-term storage of sperm, eggs and embryos. These are less costly than special farms set aside to raise non-productive animals and the technology for gene banks already exists. Sperm has already been kept frozen for some decades and embryos during the last decade. Cattle, sheep and goats have already been raised successfully from frozen embryos.

Australia has a major role to play in the preservation of the world's domestic genetic stock. We are a major source of new breeds for underdeveloped countries. We can research those countries' requirements to ensure they get the right stock with minimal effect on the genetic diversity. Genetic engineering could also be used to change genes of domestic stock.

Figure 5.7 The introduced Green Blowfly, *Lucilia cuprina*, causes much suffering to sheep through blowfly strike

Scientists have already developed methods for getting twins and quads from a single embryo so whole herds could be bred quickly to meet the needs of a developing country. As a developed country, there is less pressure on us to discard genetic diversity for economic gain.

Some people are concerned that genetic engineering is changing the natural process of evolution. They argue that we are breaking down the species concept. Some genes are not part of a species' natural gene pool. For instance, introducing a plant gene into an animal to be passed to future generations is impossible in nature. When scientists do this, we are forced to change our idea of a species.

Blow the flies

Australia has over 6000 species of flies. The most economically destructive is the introduced Green Blowfly, *Lucilia cuprina*.

Lucilia cuprina was introduced from South Africa and probably costs Australia over $50 million a year in direct losses of sheep and wool. The figure is possibly over $150 million when the costs of prevention and control are taken into account.

For sheep, blowflies mean great suffering. The flies lay eggs in the moist wool and the maggots, when they hatch, eat into the live sheep. If left unattended, the sheep will die an agonising death over many days depending on how large a maggot infestation it has.

A great deal of the grazier's time and money is lost continually checking the mobs of sheep for affected animals and then chemically treating the problem. The chemicals used do not stop the problem for long. Variation in the fly population means that the insects that survive treatment will pass their resistance to their offspring. The more potent the chemical, the greater is the natural selection for resistance and the quicker that chemical becomes useless. The short time between fly generations means that widespread resistance can be gained in a few years.

A variety of options exist. Irradiation techniques can produce fly strains with an abnormal compound chromosome that reduces fertility. When two flies of the abnormal type mate, they only produce a few offspring. When a normal fly is crossed with the abnormal strain, no offspring are produced. If large numbers of the abnormal fly are released into the wild, it is theoretically possible to replace the normal fly with the abnormal one. Within a few years the number of flies will be drastically reduced. A second option is to use the compound chromosome strain as a way of introducing into the fly population characteristics such as insecticide susceptibility.

While such options appear to be easy to create in the laboratory, there have been some problems in achieving the desired results in the wild. In 1976, hundreds of thousands of flies with a male-carried compound chromosome were released in the Canberra region. Upon sampling, scientists found that none of the 200 or so males trapped had the compound chromosome and only six of the 1200 females had one of the parents from the released strain. Possibly the wild flies were better at attracting mates than the new strain.

Further laboratory testing improved the compound strain and it was again released. This time it did establish in the wild but still with difficulty. The researchers think that the laboratory bred flies had characteristics that favoured living in the crowded laboratory situation. Another strain has been developed that is able to transfer the desired genetic characteristics to wild caught flies. Lack of funds has stopped the research.

An alternative is to genetically engineer flies with desired characteristics. The Fruit Fly, *Drosophila melanogaster*, has been well studied and its gene sequences are well known. They include a DNA fragment, known as the P-element, that has the tendency to jump from one chromosome to another. This 'jumping gene' can carry genetic information with it.

Jumping genes are pieces of DNA that are capable of moving around in the DNA from one generation to the next. When another gene is put inside a jumping gene, it becomes part of the target organism's DNA, along with the jumping gene.

The P-element can be taken from *Drosophila melanogaster* and multiplied in bacteria. It is then injected into *Lucilia cuprina* eggs, where some of it attaches to the chromosomes. In this way, certain characteristics, such as blindness, can be transferred from *Drosophila melanogaster* to *Lucilia cuprina*.

In the meantime, researchers at CSIRO's Division of Animal Production are trying another solution. In their laboratory, they are taking a plant gene that makes an anti-blowfly enzyme and placing it into sheep embryos. If the project works, the sheep will make the enzyme in their sweat glands. It will then be secreted onto the skin in the sweat. Tests have shown that the anti-blowfly enzyme does not hurt sheep tissue. If the transfer works, the animals will no longer be eaten by maggots, polluting insecticides will not harm our pastures and the sheep industry will be saved millions of dollars.

A vaccine for *Salmonella*

Each year thousands of people suffer gastroenteritis, or food poisoning. This happens after eating food contaminated with a bacteria that was identified more than a century ago as *Salmonella*. The resulting illness is called salmonellosis and it affects humans and other animals. Sheep and cattle can also suffer from the disease and meat from animals with salmonellosis poses a risk to human health.

A small team from the CSIRO Division of Animal Health has developed a live vaccine containing genetically manipulated strains of *Salmonella* for use with sheep and cattle. Researchers had already tried killed-vaccines to protect animals against salmonellosis but they were not as good as live, harmless, mutated strains of *Salmonella* bacteria. There have also been problems with the live strains. They can revert from avirulent (harmless) to virulent (harmful) strains by further mutation and so become useless as a vaccine.

The quest has been on to produce a live vaccine that did not revert to virulent, did not stay in the meat for human consumption, did not have any harmful side effects, but still provided effective long-term immunity. The answer has come from a strain of *Salmonella* called *Salmonella typhimurium*. This bacterium has a gene in its DNA,

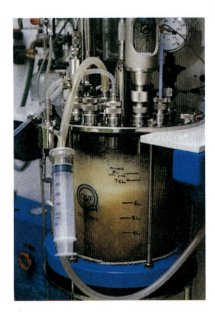

Figure 5.9 Microbes for vaccines are produced in fermenters

called *aroA*, which produces chorismic acid. This chemical gives rise to three amino acids and four aromatic compounds that are essential to the survival of the bacterium.

Two of the aromatic compounds are not found in the tissue of vertebrates and so if the bacterium cannot produce these itself, it will die in a matter of days. The *Salmonella* are therefore made harmless to sheep, cattle and humans because they will not be able to live or multiply for any length of time.

The scientists have engineered an inactive *aroA* gene by using a jumping gene. A bacteriophage is used to transfer the jumping gene from strain to strain.

In this case, the jumping gene attaches to the gene *aroA* and inactivates it. When this is achieved, the jumping gene is removed. The

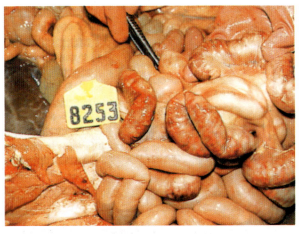

Figure 5.8 Reddening of the lower gut of a calf that died of *Salmonella* infection

60

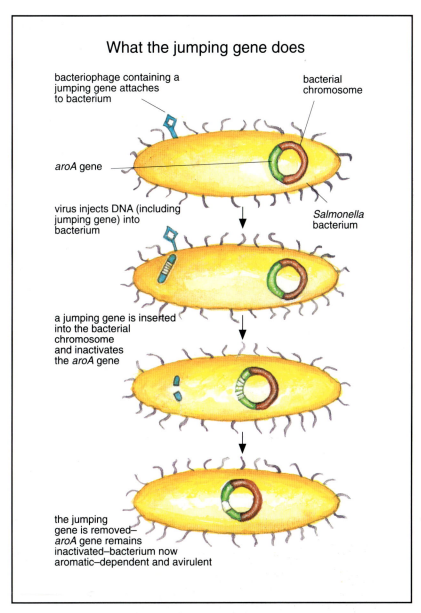

What the jumping gene does

bacteriophage containing a
jumping gene attaches
to bacterium

bacterial
chromosome

aroA gene

virus injects DNA (including
jumping gene) into
bacterium

Salmonella
bacterium

a jumping gene is inserted
into the bacterial
chromosome
and inactivates
the *aroA* gene

the jumping
gene is removed–
aroA gene remains
inactivated–bacterium now
aromatic–dependent and avirulent

Figure 5.10   Jumping genes are used to inactivate the *aroA* gene

Salmonella strain is now genetically modified. The new characteristic of being unable to make its aromatic compounds remains stable and is passed to all new generations. The new strain can then be grown in any conventional bacteriological medium. The vaccine can be given to sheep and cattle orally or by intramuscular injection and it has been shown that sheep retain immunity for six months or more.

Healthy sheep through cytokines

Most development, growth and reproduction in animals are controlled by hormones. Also affected are such diverse phenomena

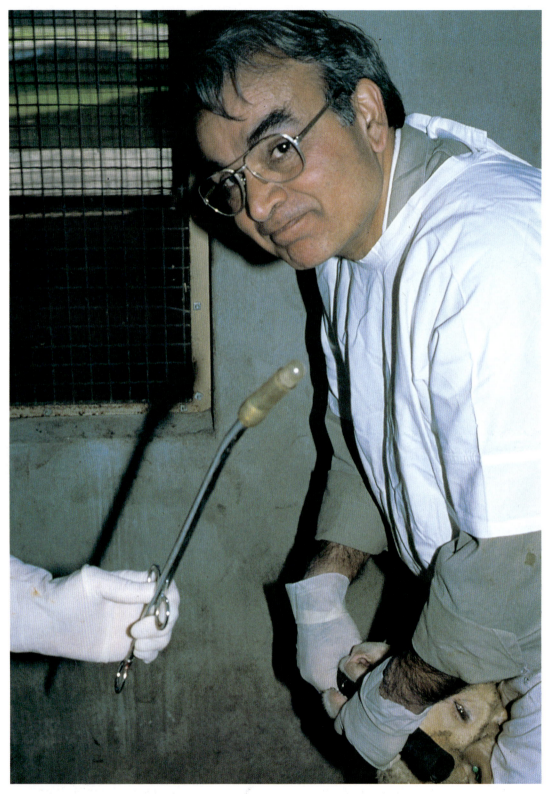

Figure 5.11   Sheep receive an oral vaccine against salmonellosis

as blood clotting, healing of wounds and response to infection. When the body responds to infection, substances called cytokines have been shown to control the response. Cytokines are a large group of proteins consisting of 100 to 200 amino acids that act either independently or in combination to perform many functions.

One type of cytokine, interferons, prevent virus infection, antibody production and activate the immune system. Another type of cytokine regulates the growth of precursor stem cells, which differentiate to become the various cells involved in combating infection, such as lymphocytes, macrophages and granulocytes. Interleukins, a third type of cytokine, regulate the action of lymphocytes in the immune response.

Cytokines are produced in small amounts. They act locally and are consumed rapidly. The same cytokine can be made by, and can act on, more than one type of cell. Some cytokines are species-specific while others work in a wide range of animals.

With the identification of cytokine gene sequences and the use of genetic engineering to make recombinant cytokine, it has been found that in many cases the effects that had been previously attributed to several factors are in fact produced by a single cytokine.

Rapid advances made in production of vaccines by genetic engineering require a detailed knowledge of the cellular reactions involved in the immune response.

Scientists at the CSIRO Division of Animal Health have genetically engineered a wide range of sheep cytokine genes. Using messenger RNA from sheep macrophages and activated T-lymphocytes, gene sequences have been amplified using a polymerase chain reaction. The products have been cloned and sequenced.

Recombinant cytokines can be used to study immune responses to infection. Such findings will assist in the identification of protective immune responses and markers for selecting disease resistant livestock.

Selecting disease resistance in sheep

Of the major sheep diseases present in Australia, worms pose particular problems because of the relationship they have evolved with their host. Internal parasites cause considerable losses in the sheep industry through reduced production and the cost of control. Current control measures involve the heavy use of chemicals. Alternatives are being sought because of the increasing cost of anti-worm treatments, the increasing numbers of parasites resistant to the chemicals and the problem of chemical residues in our food.

It was expected that vaccination would be the way to control worms. Despite much research, including using genetic engineering, no commercial vaccine has yet been found.

This has prompted a change in direction. In lambs the black scour worm, *Trichostrongylus colubriformis*, has been studied.

Fig 5.12 Healthy sheep

Some lambs seem more susceptible to this worm than others.

Controlled matings and selection has resulted in sheep that show relative resistance to *Trichostrongylus colubriformis* after vaccination with irradiated forms of the worm. The resistant lines have worm egg counts in their faeces about 90 per cent lower than the control animals. The resistance seems to continue when the lambs grow into adult sheep.

For disease resistance to be included in any selection program, it is important to have some way of identifying resistant and susceptible animals. Most work has concentrated on the Major Histocompatability Complex (MHC). MHC is a genetic system that plays an important role in many biological phenomena relating to the immune response and disease resistance. In general, the genes of MHC can be divided into several classes according to the functions they perform. Class 2 antigens are found on the surface of only a relatively few types of cells where they control things like cytokine and antibody production. The studies show that genes close to or controlling the Class 2 antigens play a role in influencing black scour worm burdens as measured by egg counts.

Animal welfare

Genetic engineering may reduce the danger of unintended genetic changes and defects which sometimes result from selective breeding. Most concerns about animals being produced with physical defects are not specific to genetic engineering and can be dealt with by animal welfare committees. The animal welfare issues of genetic engineering are looked at case-by-case and this is the role of state and local government authorities. There may even be some possibilities for increased animal welfare resulting from genetic engineering. For example, sheep will suffer less if blowfly strike can be controlled.

1. Find the following words in this chapter. What does each mean?
   - Restriction enzyme
   - Gene bank
   - Virulent
   - Killed vaccine
   - Promoter switch
2. Each paragraph in a well written passage will contain a single sentence which gives its main idea. The rest of the paragraph develops or supports the idea expressed in this key sentence. Write down the key sentence from each paragraph in this chapter. The list of key sentences is a crude summary of the chapter.
3. Compare selective breeding and genetic engineering.
4. How can genetic 'promoter switches' be used to ensure that transgenic animals do not become feral?
5. How can biotechnology help to diagnose genetic disorders?
6. Describe Pompe's disease and explain how biotechnology can contribute to its control.
7. How can blowfly strike in sheep be prevented?
8. List the steps that were taken to develop a safe vaccine against salmonellosis.
9. Trace the history of the domestication of a common farm animal.
10. What is 'biodiversity' and why is it important?
11. Paragraphs are often tied together by 'pointer words'. Pointer words are words which refer to other parts of the paragraph. They are small words such as 'he', 'she', 'it', 'they', 'this', 'each', 'who' or 'that'. Turn to the section 'Blow the flies'. Take a pencil and put a circle around each pointer word. Draw an arrow from the circle to the words to which the pointer refers.
12. 'Connectives' are words that link ideas. Connectives are joining words which show the relationship between the ideas they connect. Choose words from the following list to join the pairs of sentences below.
    - similarly
    - after
    - where
    - and
    - furthermore
    - eventually
    - meanwhile
    - but
    - although
    - often
    - because
    - a
    - which
    - or

    Write the joined sentences. You may need to change some other words to combine the sentences. You may use the same word more than once and you can use more than one word from the list in each pair of sentences. The first pair is done for you.

    Thousands of people suffer every year from food poisoning. The technical name for food poisoning is 'gastroenteritis'.
    Thousands of people suffer from gastroenteritis, or food poisoning, every year.

    The bacteria was identified more than a century ago. The bacteria is called *Salmonella.*

    The disease is called salmonellosis.
    This happens after people eat food which is contaminated with *Salmonella.*

    Sheep and cattle can suffer from salmonellosis.
    Meat from infected animals poses a risk to human health.

Chapter 6
Plants

I t is difficult to genetically engineer plants. Before it can become part of the DNA of the host, introduced DNA has to cross a cellulose cell wall, the cell membrane and then the nuclear membrane. This complex pathway presents problems to biologists.

Delivery systems needed to introduce new DNA sequences into the genome of a plant can include viruses, bacteria and physical methods. The most commonly used gene-delivery system for the plant world comes from the crown gall bacterium, *Agrobacterium tumefaciens*, which infects a wide variety of species. Part of the Ti (tumour-inducing) plasmid of this bacterium gets incorporated into the host plant's genome. The plasmid takes over the host's genetic machinery, diverting the plant's supply of the amino acid arginine into amino acids that only *Agrobacterium tumefaciens* can use. The gall that forms on the plant is outside the plant's normal controls on growth and development — just like a cancer. This improves the bacterium's success because the uncontrolled growth draws nutrients away from more productive, uninfected parts of the plant. Scientists have disarmed and modified the Ti plasmid and now use it as a method of transferring simple genes into plant cells.

Figure 6.1 *Agrobacterium tumefaciens* **normally produces tumours in plants. After the micro-organism's ability to form tumours was removed, it has been used as a 'vector' to insert foreign DNA into plant cells.**

Blue roses

Four common plants — roses, carnations, chrysanthemums and gerberas — do not have blue flowers. Through genetic engineering, scientists at the Melbourne company Calgene Pacific are trying to produce blue-flowering varieties of these plants without changing their perfume, shape, disease resistance and flower productivity.

Traditional plant breeders can look only within the same plant species for new traits to improve a flower.

However, using genetic engineering, desirable genes, such as a blue colour, may be taken from totally unrelated plants, as well as from micro-organisms and animals.

Scientists who are trying to develop a blue rose aim to isolate the 'blue' genes from plants which have blue flowers, using biochemical gene splicing techniques. After more than five years of intensive research, Calgene Pacific scientists have isolated the blue gene from the blue petunia. The selected gene will now be transferred into *Agrobacterium tumefaciens* and the Ti plasmid will carry the new blue gene into one of the chromosomes of the target plant. The transformed blue plant cells will then be grown in culture and regenerated to produce the new plant.

Calgene Pacific strengthened its position as a successful plant biotechnology company in the floriculture industry by buying its Dutch-based competitor, Florigene BV. In 1992, Florigene was granted

Figure 6.2 The blue rose is being developed by Calgene Pacific in Melbourne

permission by the Dutch regulatory authorities to produce and sell genetically modified chrysanthemums in The Netherlands. This was a 'world first' for any such plant product. Florigene now sells Europe-wide after approval by the European Economic Community in 1993.

Calgene Pacific has also developed gene-altered 'longer life' carnations, which are undergoing trials in Australia and USA. The flowers of the new varieties are claimed to have at least twice the life of conventional flowers.

Genetic engineering is now being used to control flower fragrance, development and longevity to increase our export of flowers. Once these become controllable, Australia will be able to supply improved flowers out of season to the northern hemisphere. It will also be able to directly match the needs of the Asian market. The Japanese are very fussy about colour in particular and will pay a premium for the colour of their choice. Genetic engineering techniques will make the flower business more of a fashion business. We will be able to supply 'designer flowers'.

The Australian cut flower industry is worth about $250 million a year. Of this, 10 per cent comes from exports. With specialty flowers from genetic engineering, this industry could expand greatly.

New genes for wheat

Cultivated wheat, *Triticum aestivum*, is the world's most important cereal. Its genetic history is comparatively short, going back only 12,000 years.

About that time, a new and natural cross of the many species in the Triticeae tribe began to be used. All modern wheat cultivars come from this limited genetic base of only one species.

In the early 1960s, a Japanese biologist, Professor H. Kihara, went to the Middle East to study the origins of the plants that humans first domesticated 10,000 years ago. Among the wild grasses that Professor Kihara collected were hundreds of varieties of a goat grass, closely related to wheat, known as *Triticum tauschii*. It is one of the evolutionary precursors of modern wheat. The results of this trip, as shown in the following story, show the importance of maintaining genetic biodiversity.

Genetic screening of *Triticum tauschii* began in the early 1980s by scientists in Australia, the USA and Canada. It was soon realised that the wild grass contained a vast reservoir of new genes that could potentially improve wheat. This was very exciting to the scientists because 50 per cent of the improvements in productivity have come from adding new genes into crops. To improve on this figure required the discovery of more genes.

Australia has 416 items to study: some of these are specimens from the Japanese expedition and others are from its own collecting. Of these, dozens are resistant to diseases such as wheat rust and powdery mildew. Not surprisingly, some of the Australian research has been devoted to screening the collection for salt tolerance as dryland salinity is a serious problem in many of Australia's wheat growing regions.

Work related to finding strains with resistance to cereal cyst nematode is progressing most rapidly. This nematode invades the root tips of cereal crops and costs farmers about $30 to 70 million a year. Currently, the nematode is partly controlled by rotating wheat crops with resistant crops like rapeseed or pasture legumes. Crop rotation cannot be used in all areas and it is difficult to manage.

Several Australian wheat varieties are resistant to the nematode but all are thought to get their resistance from a single gene. This spells disaster if other types of cereal cyst nematodes enter Australia. (Australia has one type but ten types have been identified overseas.)

Some varieties of *Triticum tauschii* show strong resistance to cereal cyst nematodes. Mr R. Eastwood at the Victorian Institute of Dryland Agriculture says the resistant gene (or genes) is on a section of the plant's DNA known as the 'D genome'. This is a new location for cereal cyst nematode resistance so it is likely that the gene is a completely new source of resistance.

Figure 6.3 Cotton bollworm caterpillar on a cotton boll

The D genome occurs in bread wheat. In fact, *Triticum tauschii* is the genetic ancestor of the D genome that occurs in modern plants. This makes it relatively easy to cross *Triticum tauschii* with wheat. *Triticum tauschii* has already been crossed with a variety of durum wheat called Langdon, which is susceptible to nematode attack. Results of the field trial in nematode-infested soil show that roots of the Langdon–*Triticum tauschii* cross were completely free of female nematodes while the Langdon variety averaged fifty female nematodes a plant.

Researchers at the CSIRO Division of Plant Industry have developed a genetic probe that shows the presence of the resistant gene on the D genome. This means that the gene can be 'followed' in plant-breeding research, as *Triticum tauschii* is crossed and recrossed into different wheat varieties. This will speed the research to find new resistant wheat varieties.

Insect resistant cotton

A commercial variety of cotton genetically engineered to resist a major insect pest has been developed by CSIRO scientists at the Division of Plant Industry. The new transgenic cotton plant has an extra gene inserted to make it produce a substance that is poisonous to cotton bollworms, insects that eat the plant's leaves and flower buds.

If planned field trials are a success the development may save Australia's cotton industry from possible collapse. The industry is worth about $700 million a year in export earnings. Chemicals used to control the caterpillars are becoming less effective, and their impact on the environment is being increasingly criticised.

The genetically engineered cotton variety contains a gene from *Bacillus thuringiensis*, a common bacterium. This bacterium naturally produces a protein toxic to a narrow range of insects, including cotton bollworms. As a result of the change, the leaves of the transgenic plant produce the same toxin, which kills insects eating them.

The protein is regarded as safe for fish, birds, mammals and invertebrates such as spiders (which often help control the insect pests) and has been in use for more than 20 years as a horticultural insecticide. The toxic protein gene is incorporated within the *Agrobacterium* for transfer to the cotton plant.

Fingerprinting a grapevine

Physical characteristics, such as leaf shape, number of leaves, petal shape and number, fruits and so on, have been relied on in the past to identify most plants. These are subjective measurements because the environment in which the plant is growing changes such characteristics. One plant could have been sprayed; animals may have nibbled another and changed its appearance; climate; length of growth time — all manner of factors — affect the look of a plant (the phenotype).

The only feature of the plant not affected by the plant's location and what is happening around it is its DNA structure (the genotype). Only breeding or mutation alters this. The CSIRO Division of Horticulture in

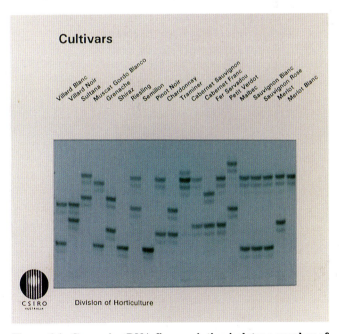

Figure 6.4 Grapevine DNA fingerprinting isolates a number of pieces of DNA, each of which shows minor differences in length between grapevine varieties

Figure 6.5   This CSIRO bred wine grape, called Tarango, has been DNA fingerprinted

Adelaide is working on an objective method for identifying plants using DNA fingerprinting (see Chapter 3).

Many horticultural crops are propagated by cloning. So, in theory, many grapevine varieties today could be the same grapes that came out of Persia about 5000 years ago. With DNA fingerprinting, it will be possible to identify whether plants have been bred true or whether they have been crossbred with others to create a new variety (or genotype). This is important to growers because clones of proven ability are the most productive and they and their products bring the highest price.

Growers whose livelihood depends on producing grapes for wine, for example, need to know they are planting, growing and harvesting vines of the best material. The only way to be nearly one hundred per cent sure is to have the plants identified by their DNA structure. Legislation now exists called Plant Variety Rights. This means dishonest growers who steal cuttings from other growers' prized

products will have a difficult time proving their ownership. The technique will probably be used for wine and table grapes, dried fruits, and other crops where identification is vital to production, processing and marketing, including hops, some flower varieties and premium fruit and nut crops.

In grapes, the process CSIRO has developed involves the isolation of a number of pieces of DNA — each of which shows minor differences in length between varieties. A chemical test precisely measures the length of these pieces of DNA. For each such piece of DNA, any one grapevine variety will contain two copies — one from the maternal and one from the paternal parent. In a photo of a DNA fingerprint for grapes, each black band represents a piece of DNA and its position on the fingerprint is a measure of its length. Fingerprints with three or four such DNA pieces have allowed identification of all grapevine varieties tested so far. At the same time, statistical tests measure the reliability of the identification.

Figure 6.6 Somacloned wheat plants can show large variation in size and ear shape. In this series, the parents are at either end and their somacloned progeny are between.

Another DNA fingerprinting identification method is based on analysis of random pieces of the plant's DNA. The technique, developed by Dupont and called Random Amplification of Polymorphic DNA (RAPD), separates many varieties of grapevines. With the development of suitable statistical methods, this technique will also be very powerful.

Somaclonal variation

According to genetic theory, when plants are reproduced from the one parent using tissue culture, all of the offspring are genetically identical. In practice, this is not true. Instead they show a wide variation that, in some cases, is greater than that shown by plants produced through crossbreeding. This variation is called somaclonal variation because the variation comes from the plant's somatic (body) cells instead of from its reproductive cells.

Why and how it happens is not yet certain. The variation can be maintained and certain characteristics such as disease resistance can be transmitted to following generations. The tissue culture technique has been around for 50 years. Plants that were not identical as required were usually thrown away without any question as to why they were different. In 1979, two scientists at CSIRO's Division of Plant Industry tried to use somaclonal variation to breed a new strain of sugar cane that is resistant to the eyespot disease *Helminthosporium sacchari*. Sugar cane is a difficult plant to breed and new strains often take ten years to produce.

It took the scientists less than one year to produce a disease resistant line and release it for testing. In their experiments, the researchers introduced a toxin made from the eyespot fungus into the newly developing cells. They found that not only did some plants have resistance to the disease but that they could be identified at that very early stage. Plants not resistant were eliminated.

Figure 6.7 Disease symptoms of sugar-cane eyespot disease produced by injecting a leaf of the susceptible parent type with fungal toxin

Figure 6.8 A somaclonal variant injected with toxin remains free of symptoms. The marks on the leaf were caused by the hypodermic needle.

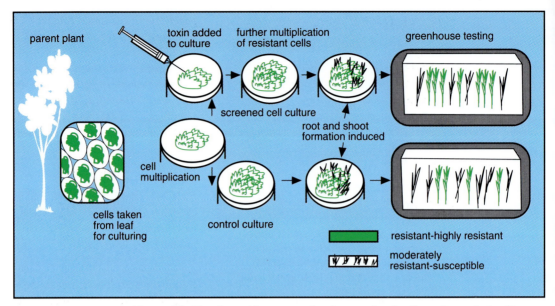

Figure 6.9 Plantlets with a range of resistance develop from the control culture. In the screened culture, the added toxin kills cells susceptible to it. As a result, a higher proportion of the plantlets show substantial resistance.

About 20 per cent of the plants showed a high resistance to the disease. The experiment then used a stronger toxin solution to see if an even greater resistance to the disease could be produced. The results confirmed that they could. A supersaturated solution of the toxin succeeded in improving the average resistance of the plants grown from the surviving cells. It was then necessary to see if this characteristic of increased resistance could be passed on from one generation to the next. It could.

Transgenic potatoes

The transgenic potato has an added gene that makes the plant resistant to potato leaf roll virus. The leaf roll virus is the most serious viral infection of potatoes world wide. Outbreaks can cause losses of up to 50 per cent in a crop. The transgenic potato has considerable potential to help the environment. At the moment, farmers spray their potatoes

up to eight times to kill the aphids that spread the leaf roll virus. The use of insecticides could be dramatically reduced.

The 'synthetic gene' is based on one taken from the virus. In the virus, the gene produces a protein that coats and protects the virus. In the potato, the gene allows the plant cell to produce the same protein. This protein protects the plant against the virus. Why this is so is not known. Nothing else in the potato seems to be affected. It has the same appearance and taste.

The transgenic potato has been approved for release by the Genetic Manipulation Advisory Committee. After five years of research, the first trial crop was picked in Queensland in mid 1993. The potatoes could be available in the market in three years if the public finds them acceptable. The first planned use is in potato chips.

River Red Gums

Many people are worried that genetically engineered life forms will become pest species. CSIRO scientists recently had success in transferring a gene from a bacterium into cells from *Eucalyptus camuldelensis,* the River Red Gum, as a first step in producing a commercial plant that will not be able to breed.

The harmless bacterial gene transferred into the gum tree is simply a marker to show that the technology works. Next, the scientists plan to repeat these experiments using two temperate eucalypts — *Eucalyptus nitens* and *Eucalyptus globulus* — both of which are important commercial plantation trees in Australia. As an environmental safety measure, the

genetically engineered plantation trees will be sterile. Once this has been achieved, there will be no risk that unwanted genes might spread into native forests. The CSIRO Gene Shears technology (see Chapter 2), which switches off unwanted genes, will be used as part of the sterility program.

Important properties, such as pest resistance and high density wood, could be changed in these commercial trees. The genetically engineered trees will not be able to breed, so they will have to be propagated by laboratory cloning. Scientists will be able to choose the best plantation specimens and produce large numbers of these special trees. Until then, the transformed trees are growing under secure laboratory conditions and it will be some time before plants are available or officially approved for field testing.

Gene mapping in trees

The forest industry is very excited about a project in the CSIRO Division of Forestry which aims to map the genes of two popular commercial species. Work has begun on mapping the DNA of *Pinus radiata* and *Eucalyptus nitens*.

The result should be a valuable new tool in tree breeding that allows much earlier selection than is possible now. With knowledge of the genetic blueprint of the two species, trees can be screened while very young for desired characteristics. Genes may also eventually be taken from one species to improve trees of another species.

The research team aims to create genetic 'linkage' maps for the two species. This involves identifying several hundred short stretches of DNA. These are called 'markers'. The location of these markers

Figure 6.10 Tissue culture, the process of regenerating whole plants from single cells, is vitally important in bio-engineering plants

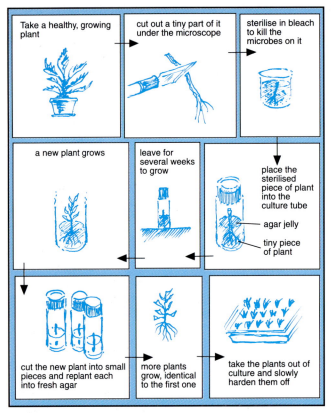

Figure 6.11   The steps in plant tissue culture

will then be worked out. The next stage will be to find correlations between the presence or absence of particular markers and characteristics of interest to forest managers, such as growth rate, pulp yield, and resistance to various diseases. Although this will not pinpoint the genes responsible for particular traits, for tree breeders, it will be just as useful. They will be able to screen young plants for the markers associated with the feature they are interested in. By doing so they will be separating out the plants with the relevant gene.

Identifying the actual genes will be the next step, but it is probably a long way off. With knowledge of the actual gene, the forest industry will be able to clone the desired genes and insert them into plants of the same species or into other species.

Who owns a life form?

Anti-genetic engineering lobbyists argue that laws relating to the technology allow businesses to own the genetic resources of the planet. The idea is that when a company patents a genetically altered organism, it is claiming ownership of another living creature. It is argued that living creatures are a resource that has always belonged in the public domain.

To counter this, the geneticists say that any time farmers pay to have their cattle artificially inseminated, they are doing the same thing. Any time people buy hybrid seeds from the nursery, they are also doing the same thing. Farmers now charge substantial amounts of money for the services or semen of stud bulls, rams and stallions. They 'own' the genetic material they have engineered by careful breeding, and they sell it at a considerable price.

Similarly, many seed companies now sell only hybrid seeds to protect their market. Plants grown from hybrid seeds cannot breed, so the customer must come back each year to the seed retailer and buy another batch. So, if you have paid money for a thoroughbred pet or for a packet of hybrid seeds, you have really paid royalties for a life form 'owned' by somebody.

Breeders have already found ways to claim fees for genetic resources, and have been doing so for generations.

1.   Find the following words in this chapter. What does each word mean?
   ● Domesticated
   ● Chromosomes

- Culture
- Somaclonal
- Toxic

2. Each paragraph in a well written passage will contain a single sentence which gives its main idea. The rest of the paragraph develops or supports the idea expressed in this key sentence. Write down the key sentence from each paragraph in this chapter. The list of key sentences is a crude summary of the chapter. Once you have produced the list, rewrite it so that it forms a precis of the chapter.

3. How can genetic engineering reduce the use of harmful pesticides?

4. How can genetically engineered varieties be stopped from becoming feral species?

5. Describe the way in which the most commonly used gene-delivery system for plants is used by genetic engineers.

6. How can biotechnology help cotton growers?

7. Why are physical characteristics not reliable for identifying grape varieties?

8. Is DNA fingerprinting a completely reliable method of identifying grape varieties? Give two reasons for your answer.

9. Discuss the issues surrounding 'Plant Variety Rights'.

10. Concerns for the social implications of technology are real and important. However, sometimes they can produce contradictory pressures. Describe one example of such a contradiction.

11. Use this chapter's subheadings to show the structure of the chapter's ideas. Use the key sentences for each paragraph to summarise the chapter. Compare this summary with the precis you prepared for question 2.

12. Paragraphs are often tied together by 'pointer words'. Pointer words are words which refer to other parts of the paragraph. They are small words such as 'he', 'she', 'it', 'they', 'this', 'each', 'who' or 'that'. Turn to the section on River Red Gums. Take a pencil and put a circle around each pointer word. Draw an arrow from the circle to the words to which the pointer refers.

13. 'Connectives' are words that link ideas. Connectives are joining words that show the relationship between the ideas they connect. Turn to the section on 'Fingerprinting a grapevine'. Using a pencil, draw a circle around each connective in the first paragraph. Draw a table which shows the connective and how it has been used. The first connective is done for you below.

Figure 6.12 Provisional Plant Variety Rights have been granted for Sunset Mandarin

Connective	Use in this sentence
such as	The following words are examples of the words before the connective.

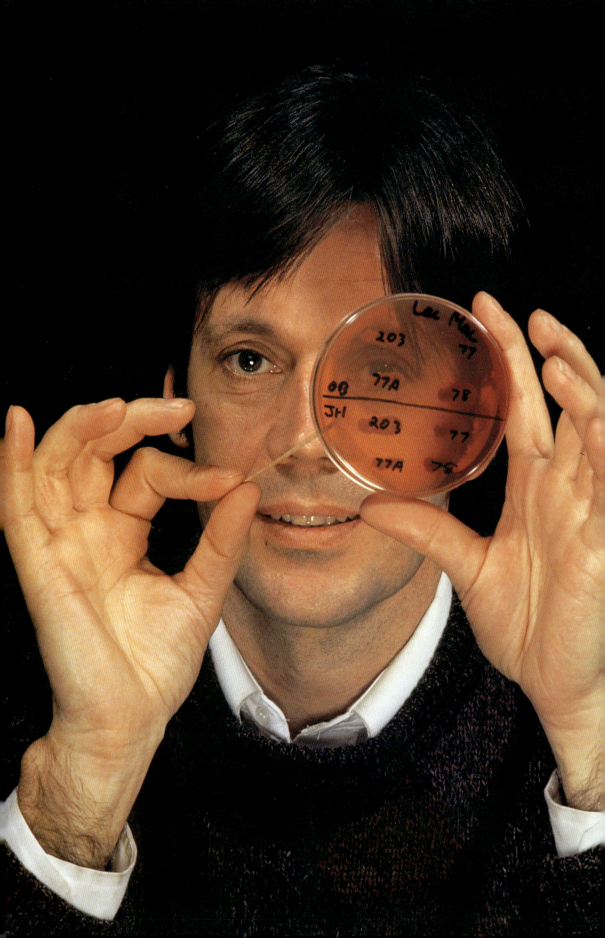

Chapter 7
The environment

S oil and water micro-organisms break down waste matter for their own use. This solves much of the world's waste management problem naturally. Now that our population has grown too much and there are more people in urban areas, the micro-organisms cannot degrade all of our wastes.

Is there a role for modern techniques of biotechnology in 'improving' on nature's techniques for cleaning up waste, making better uses of resources, or detecting pollutants? The concern of some environmentalists is that there may be dangers in changing micro-organisms in ways that will lead to unintended consequences. The scientists argue that the changes they make are tiny — one or two gene sequences out of anything up to 100,000 — and that nearly always the changes make the organisms weaker and less likely to survive in the wild. In some cases, our use of micro-organisms requires no genetic manipulation, as shown in some of these examples.

Grease-gobblers

Grease is a major environmental polluter. We release grease from wash-up water, baths, showers and industrial processes.

There are now micro-organisms that are capable of eating grease. Selective breeding and natural selection have turned normal strains of bacteria into grease-gobblers. The processes have taken years to develop. In normal circumstances, micro-organisms do not take in the complex food molecules that make up many of the greases and fats produced by human activity. The new selected and developed strains of grease-gobbling bacteria have to be able to deal with different food molecules under a range of conditions.

The bacteria, *Bacillus* spp., do not take food molecules in through their cell walls. They secrete extracellular enzymes. This means that digestion takes place outside the cell. Most enzymes will only work on one particular type of substance and under specific conditions. The scientists who produced the grease-eating micro-organisms had to first subject the organisms to unnaturally high concentrations of a pollutant such as grease or detergent. This was to produce a strain that could tolerate higher concentrations of these chemicals than found in nature. They then selected those that produced a number of different enzymes to deal with different food molecules under a variety of conditions.

Grease-eating micro-organisms have now been bred for a variety of uses. Some are being used to remove oil from polluted land, enabling the land to return to normal use. This is particularly useful around oil refineries. The micro-organisms are also good for kitchen drains, particularly in restaurants and hotels. Grease will block drains after a while if left untreated. Some factories, such as meat processing works, have an even more acute problem with large amounts of fats continually flowing down their drains. In rural areas, the grease-gobblers are ideal because there is usually no main drainage system. In septic tanks, the micro-organisms could be very useful because they can increase the natural biomass in the tank, reducing blockages and bad smells.

Waste water treatment

Each person discharges about 250 litres of water into the sewerage system each day. We produce most of the contaminated water in the home as food scraps, soaps and human waste. Micro-organisms start to break down the waste water when it enters the environment. This process uses oxygen. If the decomposition takes place too rapidly, the water may lose all of its oxygen with the resulting death of plants and animals living in the water. An important factor in assessing waste water is the strength of the waste measured as the quantity of oxygen required to decompose the waste organic matter. A biotechnology technique may be used to assess this. The amount of oxygen dissolved in the water before and after incubation in the dark at 20°C for five days measures the biological oxygen demand (BOD).

There are physical, chemical and biological treatments for waste water. The method used will depend

secondary clarifiers

aeration tanks

anaerobic contact tanks

Figure 7.1 Micro-organisms are used to treat sewage in this Victorian plant

on what is in the water, where it is and how much there is. A combination of methods may be used. Biological treatment of fine solids and dissolved organic material uses biotechnology for primary, secondary and tertiary treatments. Both primary and secondary treatments occur in most cities in the developed world. In some large cities however, Sydney for example, only primary treatment occurs. Normal sewage usually does not undergo tertiary treatment.

Primary treatment removes the most easily separated contaminants. These include oil films, particles that float and particles that settle easily. Primary treatment removes about 60 to 70 per cent of the suspended matter as sludge. The water remaining after primary treatment usually only goes into the environment if there are already large volumes of water there to dilute it. For example, primary-treated sewage empties into the sea in Sydney. We now realise that the sea is not diluting the sewage enough.

Aerobic and anaerobic biological processes occur in secondary treatment. Aerobic treatment requires both an active micro-organism population and oxygen. Temperature, pH, inorganic nutrients and restricted levels of toxins are also necessary. There are several methods for secondary treatment. One of the oldest is to simply spread the material over an area of land. This method is not suitable for the large quantities of waste produced by cities. Trickle filters are an alternative. These rely on trickling effluent gradually through a bed of stones or other material with a large surface area. With added oxygen, micro-organisms on the rocks consume any fine sediment and organic particles. In

large water treatment systems, rotating biological filters use this principle. Here, the micro-organisms are grown on plates mounted on a shaft that slowly rotates through the water being treated.

The activated-sludge process is another method. It mixes a micro-organism culture with the water. Oxygen bubbles through the mixture. After a time, the water is filtered. This settles out the activated-sludge and the waste material it has removed from the water.

The sludge from both the primary and the secondary treatment of water is anaerobically processed. There is no oxygen with anaerobic treatment. This means that a different group of micro-organisms breaks down the material into simpler compounds such as methane, carbon dioxide and hydrogen. This is the process that takes place on a small scale in a septic tank. On a large scale, the ponds are up to 4 metres deep.

Tertiary treatment of waste water is not usual. Treatments include one or more of a number of processes such as electrodialysis, reverse osmosis, deep-bed filtration and absorption.

Mineral recovery

Micro-organisms exist that will leach minerals from rocks (that is, separate a mineral from the ore-containing rock) by simply dissolving the mineral and absorbing it. The micro-organisms are then recovered along with the valuable mineral.

Metals are often present in rock as insoluble metal sulfides, such as lead sulfide. Conventional methods for extracting the metal from the mined ore require high-temperature smelting processes that decompose the ore to metal and sulfur dioxide. The crude metal can then be refined, possibly by electrolysis. These processes use considerable amounts of energy and produce significant pollution. It is now possible to extract the metal using bacteria.

In 1947, the bacterium *Thiobacillus ferrooxidans* was shown to cause the acid pollution associated with coal mining. The bacterium was decomposing iron sulfide into sulfuric acid and iron. The bacterium is able to get the energy it needs from the oxidation of the sulfide and it gets the carbon necessary for making its organo-carbon compounds from atmospheric carbon dioxide.

Two methods of sulfide oxidation have now been identified.

1 Direct bacterial leaching: The bacterium attacks those mineral components susceptible to oxidation. The bacterial enzymes catalyse reactions which release electrons from the sulfur or iron. The electrons are transported to the bacterium where they combine with oxygen to form water. Energy is released as the electrons transfer to the bacterium and the bacterium can use this energy. The metal ions never enter the bacterium.

2 Indirect bacterial leaching: This does not involve enzyme activity. Bacteria attack soluble ferrous ions and in so doing produce ferric ions. Ferric ions are powerful oxidising agents and transform metal sulfides into metal ions and sulfuric acid. This reaction regenerates the ferrous ions and the cycle continues. This method is widespread because iron deposits often contain metals such as copper and uranium.

Extracting oil

Oil does not exist in a large cave underground. It is not just a matter of finding a pool and pumping it out. The oil is held in the very small spaces between the sand grains that make up the rock. It is similar to water held in a sponge. Miners using conventional methods do not have all the oil just because it stops flowing.

On average, about 70 per cent of the oil is left in the ground and cannot be recovered with traditional techniques. A thin film of oil held by surface tension coats the surfaces of the sand grains. Various means have been tried, with different degrees of success, for trying to recover more oil. They aim to maintain reservoir pressure by injecting water or gas down a hole drilled parallel to the production well. Chemical 'surfactants' that lower the surface tension of the oil have also been tried.

CSIRO tried to find a bacterium to extract the oil left when the well stops producing. Researchers looked at the nutritional requirements of naturally occurring micro-organisms from oil wells. They developed a nutrient cocktail that prompted the bacteria to produce a particular end-product. This end-product was a surfactant that could be easily extracted from the well. The idea

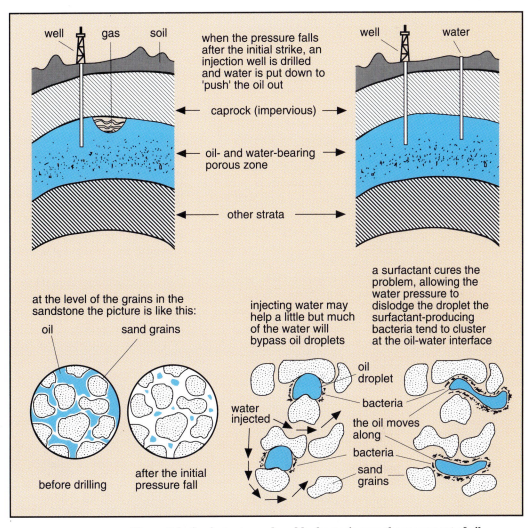

Figure 7.2   Surfactants produced by bacteria ease the movement of oil

was to provide a nutrient that would cause the bacteria to produce other products. If the bacteria produced gas, this would provide pressure to move the oil products. The bacteria might also produce a substance to improve permeability of the rock around the well. The chosen nutrient was pumped down the well. The bacteria already living increased in number and produced the products required. When oil extraction stopped and the nutrient was no longer pumped down, the bacteria returned to their original population size and function.

This process was tried in Australia, in the Queensland Alton Field. Each well is different so it took about three months to analyse the bacterial needs of a well. The Alton Field began production in 1966 and quickly started to decline about 15 per cent each year. Production increased 50 per cent a month after treatment with bacteria. The level was maintained for the 12 months following. Bacteria increased from 1000 per millilitre to 100,000 per millilitre and the volume of oil recovered increased. The procedure only worked when the temperature of the oil reservoir was suitable for bacterial activity. The bacterial extraction method is not being used any more because it is still cheaper for oil companies to look for new wells.

Toxic wastes

There will probably be some oil-digesting bacteria already present at the site of an accidental oil spill. For years, contaminated areas have been seeded with additional oil-digesting bacteria. The bacteria oxidise the oil — a compound rich in carbon — to satisfy their energy requirements. However, like all living things, bacteria need nitrogen, phosphorus, and other elements for growth. It may be necessary, therefore, to add certain nutrients to maintain the correct carbon:nitrogen ratio to ensure the greatest level of reproduction by the bacteria.

Currently, polluted land sites are often excavated and the contaminated material is either buried as landfill (which merely transfers the pollution to a second site) or incinerated (which is costly and may produce toxic ash and fumes). Chemical treatment is a possibility but it may also be polluting, or the problem can be left where it is. In that case, attempts would have to be made to prevent any spread or contamination of groundwater.

Figure 7.3 **Scientists watch mixing of the nutrient solution that feeds the bacteria down the well**

Bioremediation is the use of micro-organisms to treat contaminated sites. It is a cheap and efficient way of removing toxic chemicals. The choice of micro-organisms, and the method used to apply them, depends on the type and concentration of the contaminant and the soil characteristics. The availability of oxygen, water, and nutrients or unfavourable pH or temperature limits the rate of degradation. Careful monitoring of each treatment site is needed. In Canberra, a research team has developed a method of changing the suite of micro-organisms throughout the breakdown process, leading to a more rapid and complete destruction of the contaminant.

Rabbits and foxes

The European rabbit is Australia's most serious vertebrate pest. Some 200 to 300 million rabbits cause a $90 million loss in agricultural production each year, and foxes wreak havoc on native wildlife and new-born lambs.

To 'rid Australia of this scourge', the Intercolonial Rabbit Commission was set up 100 years ago. All efforts have failed — rabbits are still a big problem. The myxoma virus, released by CSIRO 40 years ago, has been the most effective weapon against them. Due to the virus, there are fewer rabbits now in many parts of the country, but in other parts, particularly drier areas, rabbit numbers are still very high.

The Vertebrate Biocontrol Centre in Canberra is now looking at ways of changing the myxoma virus so that it will carry a gene to disrupt the breeding cycle of the female rabbit. A similar method with another virus is planned for the fox. This is the first attempt to attack the rabbit and fox problems by regulating birth rather than death.

Figure 7.4   A feral rabbit—introduced to Australia from Europe

Figure 7.5 Rabbits head Australia's list of vertebrate pests because of the amount of widespread land degradation they cause

The plan is to cause the rabbit's immune system to attack the reproductive cells of its own species. Sperm and egg will not recognise each other! Directly under attack with this approach is the high reproductive capacity of the rabbit, the feature that gives the animal its great advantage. However, it is important to do this without disturbing the distinct social hierarchy of rabbits, and foxes too. In each species, dominant females are the main breeders. Making them infertile will permanently affect the population. Other methods that attempt to control by killing off only result in subordinate animals replacing the breeders and quickly rebuilding the population.

In its transgenic form, the virus would no longer kill by causing myxomatosis, but would be used simply as a vehicle to carry new genes into rabbit cells. The myxoma virus remains the ideal control agent because it is specific to rabbits.

The virus' genetic blueprint is under analysis and the search is on for suitable sites to insert new genes. The virus has to be able to carry an extra gene without losing its ability to infect rabbits. Fortunately, the myxoma virus does have the right structure.

The second phase of the program is to find a protein that is part of fertilisation in rabbits, isolate the gene responsible, clone it and add it to the virus. On the surface of the sperm there are many proteins involved with the sperm making its first contact with the egg, burrowing through the egg coat, and recognising and binding to the membranes of the egg. Scientists can identify these proteins now and synthesise the genes for them, using molecular biology techniques.

The plan is simple: the myxoma virus receives a gene for a protein that is part of rabbit fertilisation; the virus infects the animal; the virus gets the cells to make the sexually significant protein; the rabbit then makes antibodies to eliminate it; these remain in the system and attack the protein whenever it occurs in the sexual cycle.

So far, the researchers have isolated a useful protein, part of the coating on the sperm's tail, and identified the encoding gene. The next step is to splice the tail protein gene into the virus.

Biosensors

Humans have a very poor sense of smell. This is why we use a number of other animals, such as dogs, to detect a variety of substances. CSIRO researchers are developing a new generation of biosensors that are more efficient than the noses of many animals.

Their aim is to produce a battery-powered sensor about the size of a pocket calculator that can detect a number of specific odours. The detecting part will be a series of wafers, where each wafer is an ultrasonic detector coated with a single molecular layer of monoclonal antibodies (see Chapter 3). Each monoclonal antibody responds to the presence of only one chemical at concentrations of only a few parts per billion.

To sniff out odours at low concentrations, an operator will pass the machine over an object. The machine will contain a wafer appropriate for the substance to be detected. There are many uses for such a machine. For instance, it may detect hydrogen sulfide leaking from sewage treatment works, drugs in luggage, off-odours in a food container, pollutants from an industrial process or measure the amounts of sweeteners and additives in foods.

The team is now working on techniques to produce monoclonal antibodies on a commercial scale since there is an international market for biosensors.

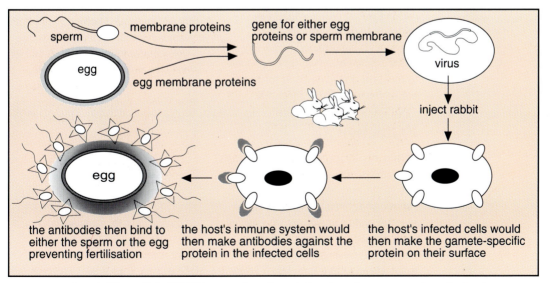

the antibodies then bind to either the sperm or the egg preventing fertilisation

the host's immune system would then make antibodies against the protein in the infected cells

the host's infected cells would then make the gamete-specific protein on their surface

Figure 7.6   CSIRO researchers hope to 'engineer' a myxoma virus strain carrying rabbit genes that will make infected rabbits infertile

Future factories

One of the most novel applications of plant genetic engineering could be to modify oil-producing plants, such as sunflower and canola, to make a wide range of industrial lubricants, detergents and cosmetic compounds. Biodegradable packaging materials may also be produced with gene technologies.

We use petrochemicals to produce plastics now. In the future we may use starch-yielding plants such as wheat and potatoes to produce particular kinds of starches to meet the same industrial purposes. In effect, plants may become biological factories making specialty chemicals. If it works, our factories won't have to rely on the mining of non-renewable resources. Our farms would not have any difficulty in becoming profitable if they were the suppliers of such necessary raw materials for industry.

The concept of biological factories could extend to transgenic animals. One area of research, for example, is making rare proteins which can be used for pharmaceuticals. For example, it is predicted that a few transgenic animals would be enough to meet the world demand for the human blood clotting factor IX.

Extinction

In the movie 'Jurassic Park', scientists were able to recreate dinosaurs. They did this after reading the dinosaurs' genetic code from samples of their blood found in mosquitoes preserved in amber. While 'Jurassic Park' is fiction, scientists are looking at DNA in fossils and have samples of DNA from recently extinct animals.

The Tasmanian Tiger, or Thylacine, was a pouched, wolf-sized animal that once roamed Australia. About 1000 years ago, Thylacines disappeared from the mainland, probably because of competition from dingoes. However, they survived in Tasmania, but only until European settlers and their sheep arrived. Thylacines killed the sheep and so the settlers were paid a bounty for the skins of any Thylacines they shot. The number of Thylacines dwindled until the last one died in Hobart Zoo in 1936. The Australian Museum in Sydney has a baby Thylacine from the 1860s preserved in a bottle of alcohol. Perhaps it is possible in the distant future to use the DNA to bring the Thylacine back to life.

DNA has been extracted from a number of different fossils. It has been taken from weevil-like insects preserved in amber for 120 million years, the bones of North American horse fossils 2.5 million years old, and tiny scraps of hair, tissue and blood clinging to stone tools thousands of years old. The DNA in this last example is probably from animals hunted, such as the extinct woolly mammoth. This DNA can be checked against other mammoth DNA taken from frozen animals preserved in ice.

DNA has also been extracted from extinct Australian marsupials such as diprotodonts, which are thought to have been extinct for 20,000 years. The DNA is extracted from ground-up bones. It is then analysed using genetic techniques. Even older Australian fossils have yielded DNA. One day we may wish to reconstruct the paddle-footed marine animal called a plesiosaur. DNA has been taken from the fossilised bones of this animal which lived 160 million years ago.

To use the DNA to reproduce these animals, it would be necessary to have the full sequence of the DNA code. While this is not available for the older animals, in the case of the Thylacine it is. Once the code is known, scientists will have to use that code to make DNA molecules billions of characters long. The best they can do at the moment is a few hundred characters. With that all done, the DNA would then need to be put into a living embryo. No one is sure whether that can be done.

To resurrect extinct animals raises important ethical questions. If we have the technology, will we fight less to save animals, plants, other organisms and ecosystems from extinction? Is it right to resurrect an animal which our ancestors sent to extinction? Could we also resurrect entire ecosystems so that the organism had a chance to survive in the 'wild'? Maybe the climate has changed. Where should we stop?

1. Find the following words in this chapter. What does each word mean?
   ● Natural selection
   ● Sewerage ● Sewage
   ● pH ● Crude
   ● Anaerobic ● Surfactant
2. Prepare a point form summary of this chapter and then use it as the basis for a precis.
3. How are human wastes broken down in nature? Why doesn't this system work in large human communities?
4. Why is the use of grease-eating bacteria an example of biotechnology?
5. Draw a flow chart to show the various stages for treatment of waste water.

6. Explain why coal waste is such a serious pollution issue.
7. Explain why causing sterility in rabbits is likely to reduce rabbit numbers more than killing adult rabbits.
8. CSIRO has developed bacterial methods for maintaining oil production. These methods are not used commercially. Explain why this is so and discuss the implications of such decisions.
9. Research the various methods for controlling rabbits that have been attempted in Australia.
10. Research the use of biosensors in industry.
12. Paragraphs are often tied together by 'pointer words'. Pointer words are words which refer to other parts of the paragraph. They are small words such as 'he', 'she', 'it', 'they', 'this', 'each', 'who' or 'that'. Turn to the section on waste water treatment. Take a pencil and put a circle around each pointer word. Draw an arrow from the circle to the words to which the pointer refers.
13. 'Connectives' are words that link ideas. Connectives are joining words which show the relationship between the ideas they connect. Choose words from the following list to join the sentences in the first two paragraphs of the section on 'Grease-gobblers'.
    ● similarly ● after
    ● where ● and
    ● furthermore ● eventually
    ● meanwhile ● but
    ● although ● often
    ● because ● as
    ● which ● or
    Write the joined sentences.
    You may need to change some other words to combine the sentences. You may use the same word more than once and you can use more than one word from the list.

Index

RNA 18, 19
  see also rRNA *and* mRNA

Salmonella vaccine 59–61, 62
selective breeding 1–2
sewage treatment 80–1
sheep
    disease resistance 63–4
    health 61, 63
SIROCURD process 43–5
social implications, of gene technology
    7–8, 32
somaclonal variation 73–4
somatic gene therapy 29
species concept 58
sugar cane, resistance to eyespot
disease 73–4
sulfide oxidation 82
surfactants 82, 83

Tasmanian Tiger 88
Thiobacillus ferrooxidans 82
Thylacine 88–9
tissue culture 75, 76
tomato ripening 46, 48

toxic wastes 84–5
transforming animals 51–2
transgenics 15, 46, 48
    animals 88
    mice 52–3
    plants 3, 4
    potatoes 47, 74
    tomatoes 46, 48
translation 14–15
Triticum tauschii 69, 70
tRNA 15, *see also* RNA
tropical cattle breeding 55, 57

United Nations 8
ultrafiltration 44

vaccines 23–4, 59–61
    killed, 23, 60
    live, 23, 24, 60
viruses 19, 85, 86–7

waste management 79–81
waste water treatment 80–1
wheat genes 69–70
whey 35